Guide to Venomous and Medically Important Invertebrates

David E. Bowles, James A. Swaby and Harold J. Harlan

We dedicate this book to our families who supported us, our colleagues who encouraged us, and to all those travellers and outdoor adventures who inspired us to write this book.

Guide to Venomous and Medically Important Invertebrates

David E. Bowles, James A. Swaby and Harold J. Harlan

CSIRO

PUBLISHING

A catalogue record for this book is available from the National Library of Australia.

Published by:

CSIRO Publishing
36 Gardiner Road, Clayton VIC 3168
Private Bag 10, Clayton South VIC 3169
Australia

Telephone: +61 3 9545 8400
Email: publishing.sales@csiro.au
Website: www.publish.csiro.au

Front cover: (top, left to right) Bulldog ant (Marshall Hedlin/Flickr); *Culex annulirostris* (Stephen Doggett); Saddleback caterpillar (Katja Schultz/Flickr). (bottom, clockwise from left) East coast sea nettle (Antoine Taveneaux); Chigger bites on lower leg (Elin/Flickr); European house dust mites (Gilles San Martin/Wikipedia); Tsetse fly (Oregon State University/Flickr)
Back cover: (left to right) Asian predatory wasp (Gilles San Martin/Flickr); Assasin bug (James Niland/Flickr); Bearded fire worm (Prilfish/Flickr)

Set in 9.5/12 Minion Pro
Edited by Peter Storer
Cover design by Andrew Weatherill
Typeset by Desktop Concepts Pty Ltd, Melbourne
Index by Bruce Gillespie
Printed by Ingram Lightning Source

CSIRO Publishing publishes and distributes scientific, technical and health science books, magazines and journals from Australia to a worldwide audience and conducts these activities autonomously from the research activities of the Commonwealth Scientific and Industrial Research Organisation (CSIRO). The views expressed in this publication are those of the author(s) and do not necessarily represent those of, and should not be attributed to, the publisher or CSIRO. The copyright owner shall not be liable for technical or other errors or omissions contained herein. The reader/user accepts all risks and responsibility for losses, damages, costs and other consequences resulting directly or indirectly from using this information.

Oct25_RP_ILS

Contents

Preface

Whether you are a world traveller, an outdoor adventurer or a recreationist out for a day trip, you face an array of potential threats from invertebrates. We envision two primary uses for this work. It can serve as a traveller's reference to assess the threats that might be present in a region to which you might travel. It also can serve as guide for medical professionals, and lay public to make preliminary assessments as to whether or not a particular invertebrate attack poses any threats, and how to mitigate those threats. This book does not deal with helminth worms and other internal parasites because doing so would make this work impractically large. We also do not attempt to cover the full range of control or management options for invertebrate pests because that is a vast topic in its own right and beyond the scope of this book.

This guide presents a basic yet sound understanding of potentially dangerous invertebrates that travellers and adventures may encounter. Our intent is not to provide a definitive or exhaustive description of every species, but instead we briefly discuss biological, physical and behavioural characteristics of key groups in an attempt to give the reader a better direction on identity. We have attempted to show in photographs the appearance of the animals we discuss, but we recognise that the range of variation seen in nature may not be entirely reflected in the photos we present. Distributional maps are presented only as general information and are not intended to be specific range boundaries.

This guide considers only animals posing a direct contact threat, not toxic food reactions, contact allergies or poisonous animals that must be consumed. In addition, a few other invertebrates are included because they are often incorrectly presumed dangerous. We reviewed and consolidated the exhaustive literature available, but, to make it concise, individual information sources are not cited in the text. We do list selected reliable resources and references should the reader be interested in learning more about the animals described.

To make the guide easy to use while 'on the go', we streamlined the text and used common language to describe the invertebrates and their associated reaction symptoms. In some situations, we used medical and technical terms for brevity or to avoid confusion. Similarly, many invertebrates are commonly called different names, depending on location. We include those common names, but associate them and all other information with their presently accepted scientific name and associated higher taxonomic nomenclature (phylum, class, order, family, genus). This will assist you to find additional information on those species or groups.

Most medical treatments included in this book involve only basic first-aid procedures, although some responses may involve complex medial intervention. Treatment guidelines, when presented, are based on information available when this guide was written. Travellers/adventures and medical-care providers should consult the most current, reliable information available to ensure correct treatment regimens. Medical practitioners are solely responsible for medications, dosages and therapies prescribed.

About the authors

Dr David E. Bowles holds a BS in Biology and Natural Resources from Ball State University, a MS in Aquatic Biology from Southwest Texas State University, and a PhD in Entomology from the University of Arkansas, Fayetteville. He also is a graduate of the United States Air Force (USAF) Squadron Officer School, the Air Command and Staff College, and Air War College. Dr Bowles retired as a USAF Reserve Colonel in 2015 following a career of active duty and reserve assignments. He worked in various medical entomology and intelligence assignments located at the USAF School of Aerospace Medicine, USAF Armstrong Laboratory, USAF Force Protection Battlelab, USAF Institute of Operational Health, USAF Research Laboratory, National Center for Medical Intelligence and Armed Forces Pest Management Board. He also formerly served as the USAF Command Pest Management Professional for Air University Command and the USAF Surgeon General's representative to the Armed Forces Pest Management Board. He is a veteran of operations Desert Storm, Iraqi Freedom and Enduring Freedom. He currently serves as Adjunct Professor of Biology at Missouri State University. Dr Bowles has published extensively on a variety of subjects including taxonomy and ecology of aquatic insects and Crustacea, aquatic botany, fisheries biology and medical entomology.

Dr James A. Swaby holds a BS in Biology from Concordia College, Moorehead, Minnesota; and MS and PhD in Entomology from Oregon State University, Corvallis, Oregon. He eventually achieved four Entomological Society of America Board Certifications (BCE) in Medical/Veterinary Entomology, Urban and Industrial Entomology, Plant Related Entomology, and General Entomology. He joined the United States Air Force (AF) in 1982, and retired as a Colonel in 2012. But his professional entomology career spans over 40 year and includes research, teaching, publishing, and consulting on pest management, vector-borne and zoonotic diseases, and related biological factors impacting human health. Those issues took him to 27 countries, 41 US states and over 200 trouble spots. Of the 14 job titles he held, those pertinent to this guide include Officer in Charge (OIC) of the AF Vector Surveillance Program, Plague Surveillance Program OIC, Dengue Surveillance Program OIC and the Principal Investigator on the AF Environmental, Vector-borne, and Zoonotic Disease Surveillance Initiative to develop advanced surveillance methods for remote, austere locations. He rose through the AF ranks to the Medical Entomology career field manager as the Chief Consultant for Medical Entomology and as the Consultant for Medical Entomology to the Air Surgeon General. His academic appointments included: AF School of Aerospace Medicine Operational Entomology Course Developer and Course Supervisor; USAF Master Training Instructor; USAF Academy Associate

Professor of Biology; Defense Medical Readiness Training Institute Affiliate Instructor; and Graduate Faculty, University of Texas Health Sciences Center, San Antonio, Texas. He also served as National Institutes of Health's National Advisory Research Resource Council Ex Officio Voting Member and on the University of Texas Institute for Integration of Medicine and Science's Clinical & Translational Science Award Working Group. He is an AF Squadron Officer School, Air Command and Staff College, and Air War College graduate, Gulf War veteran, and received numerous military, academic and research awards. He continues to volunteer: consulting, writing and lecturing on medical entomology and public health.

Dr Harold J. Harlan earned his BS, MS, and PhD degrees from Ohio State University, and specialises in medical entomology and acarology. He served 25 years on active duty as a USA Army Medical Entomologist, retiring in 1994. He was Assistant Professor, Uniformed Services University of the Health Sciences, 1990–1994, and Adjunct Professor, providing invited lectures and laboratory sessions in their Tropical Medicine Courses annually 1995–2017. He has been a Board Certified Entomologist (BCE) in Medical and Veterinary Entomology since 1973, and in Urban and Industrial Entomology since 1999. Dr Harlan has published more than 40 articles in scientifically peer-reviewed journals or book chapters, and he is a recognised world authority on bed bugs. He was the National Pest Management Association staff entomologist for 9 years, and a Senior Entomologist for the Armed Forces Pest Management Board (AFPMB) for 8 years. While at the AFPMB, he gathered and developed the bulk of text content in their Living Hazards Database (LHD). Two arthropod species new to science were named in his honour, and he was a member of MENSA's Maryland Chapter, 1995–2000. He has been an invited speaker to multiple medical entomology, public health, and pest management professional audiences in many states, as well as in Belize, Canada, Dubai, Honduras, Hong Kong, Puerto Rico and Panama, addressing such subjects as mosquito biology and management, bed bugs and hazardous biota (including hazardous arthropods, invertebrates and snakes). He retired from full-time employment in October 2013. He has remained professionally active since, giving free public presentations on various critters (upon request), doing limited consulting on pests and pest-generated allergens, and helping revise the B.C.E Core and the B.C.E. Medical/Veterinary qualifying exams.

Acknowledgements

We thank our many civilian and military colleagues, retired and active, for their comradery, professionalism and assistance over our collective careers. This guide would not have been possible without you. Photograph permissions were kindly given by Jim Occi (Center for Vector Biology), Bernard Picton (National Museum of Ireland) and Heather Stockdale Walden (University of Florida). We especially thank Stephen L. Doggett (University of Sydney) and James Gathany (US Centers for Disease Control and Prevention) for there generous donations of superb photographs. We also greatly appreciate our families who graciously allowed us time to work on this guide and otherwise supported our respective careers and world travels.

Dangerous invertebrates

Invertebrates are animals that lack a spinal column, and they include a vast diversity of body types and sizes. The invertebrates include the sponges, coelenterates (jellyfish, corals and relatives), non-parasitic worms, echinoderms (e.g. sea stars, sea urchins), molluscs (e.g. clams and snails), crustaceans and arthropods (scorpions, spiders, ticks, mites, insects, centipedes and millipedes). Regardless of where you travel, even if you are in your own backyard, you are likely to encounter invertebrates. Although most of them are completely harmless and beneficial, many people are squeamish or fearful about seeing or contacting invertebrates. Some invertebrates are medically important. They can annoy, cause pain through bites and stings, transmit various diseases and potentially cause death. For those species, it is important to be able to recognise them, understand potential risks, learn how to avoid unwanted contact, and properly respond to injuries or illness.

Arthropods (Phylum: Arthropoda) are, by far, the invertebrates most likely to cause human suffering, illness or death. The arthropods share the common features of having an exoskeleton, segmented bodies with paired, segmented appendages and bilateral symmetry. Included here are those arthropods that are medically important or can inflict painful injuries. They include representatives from among the insects, arachnids, centipedes and millipedes.

The two most prominent medically important arthropod groups are the arachnids and insects (Classes: Arachnida and Insecta, respectively). Insects as a group have been, and continue to be, responsible for much human suffering and mortality, although relatively few species are involved. Arachnids can induce considerable fear and loathing among many people, but, similar to insects, most pose no threat to people. Other arthropods, such as the millipedes and centipedes, have only a few medically important representatives.

The arachnids include the spiders, scorpions, ticks and mites. Various arachnids can affect your health via direct injury (bites and stings), envenomation, disease transmission and allergic reactions. Globally, a few spiders and scorpions can cause significant medical illness through envenomation. Scorpion stings represent the most important arachnid envenomation source throughout the world, causing considerable morbidity among adults and numerous deaths among children. However, only a few species (primarily distributed in Africa, the Middle East and Latin America) can often inflict lethal stings. Widow spiders (*Latrodectus* spp.), recluse spiders (*Loxosceles* spp.), the Australian funnel web spiders (*Atrax* and *Hadronyche* spp.), and the Brazilian wandering or banana spiders (*Phoneutria* spp.) are among the most medically important spiders. Ticks and mites are capable of transmitting serious, sometimes fatal,

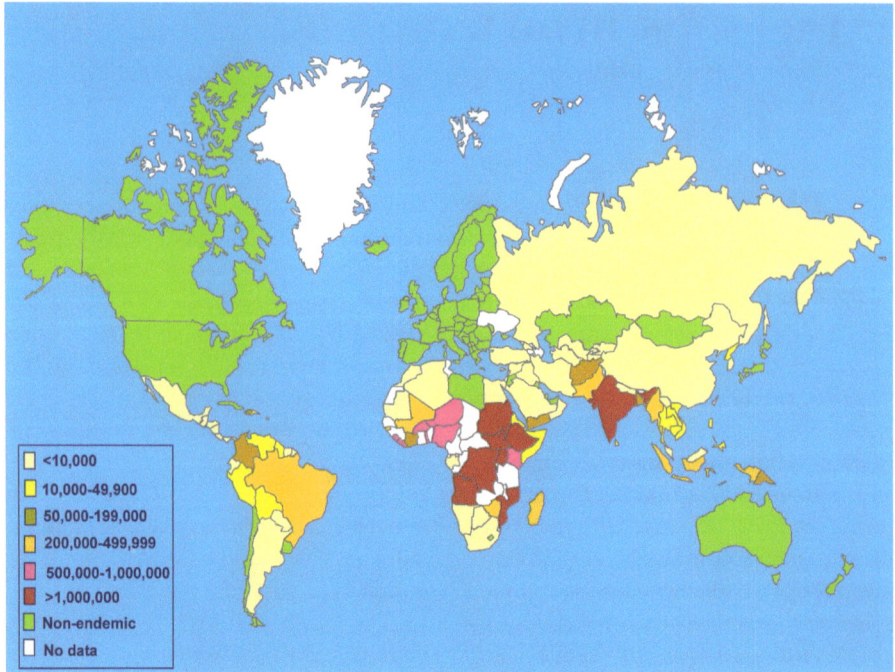

☐	<10,000
☐	10,000–49,900
☐	50,000–199,000
☐	200,000–499,999
☐	500,000–1,000,000
☐	>1,000,000
☐	Non-endemic
☐	No data

Malaria cases worldwide. Source: Adapted from the World Health Organization

diseases, in addition to causing irritation and annoyance. Indeed, some tick-borne diseases have dangerously high fatality rates. Finally, some arachnids (e.g. camel spiders) appear frightening, which may cause psychological distress, but are harmless if left alone.

Of over a million estimated species of insects worldwide, literally thousands of species can potentially affect human health and comfort through painful stings, bites and transmitting diseases. Here, we consider these medically important insect groups: lice, cockroaches, true bugs (Hemipterans), Hymenoptera (bees, ants and wasps), moths, beetles, flies and fleas. Mosquitoes are responsible for transmitting most arthropod-borne diseases. Historically, malaria, yellow fever and dengue have claimed the lives of millions of people. All three diseases remain significant threats to human health. Malaria is endemic in over 90 countries, with 500 million new cases diagnosed yearly, with nearly 3 million fatalities.

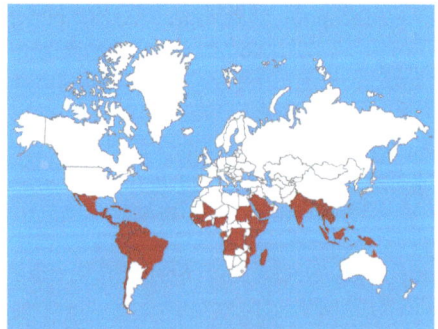

Dengue cases worldwide. Source: Adapted from the World Health Organization

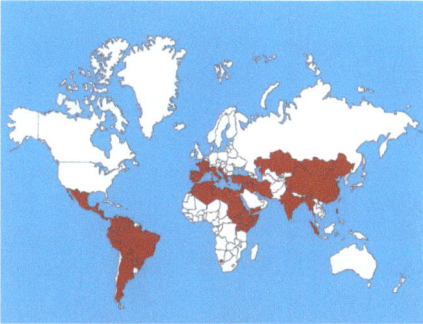

Leishmaniasis cases worldwide. Source: Adapted from the World Health Organization

Dengue is the world's most important mosquito-borne viral disease. Millions of people are at risk of infection worldwide and 20 million new cases reported yearly from over 100 countries. In 1995, the worst dengue epidemic in Latin America and the Caribbean occurred across 14 countries, causing more than 200 000 dengue fever cases and almost 6000 cases of dengue haemorrhagic fever.

Many major cities, especially in the Americas, have persistent populations of the mosquitoes that transmit yellow fever, and they are at risk for potentially devastating epidemics of this disease. Another mosquito-borne parasitic disease, lymphatic filariasis (often causing elephantiasis), infects ~120 million people in tropical Africa, India, South-East Asia, the Pacific Islands, and South and Central America.

Insects transmit numerous other parasites and disease agents. Sand flies transmit leishmaniasis, with roughly 350 million people at risk where these parasites occur.

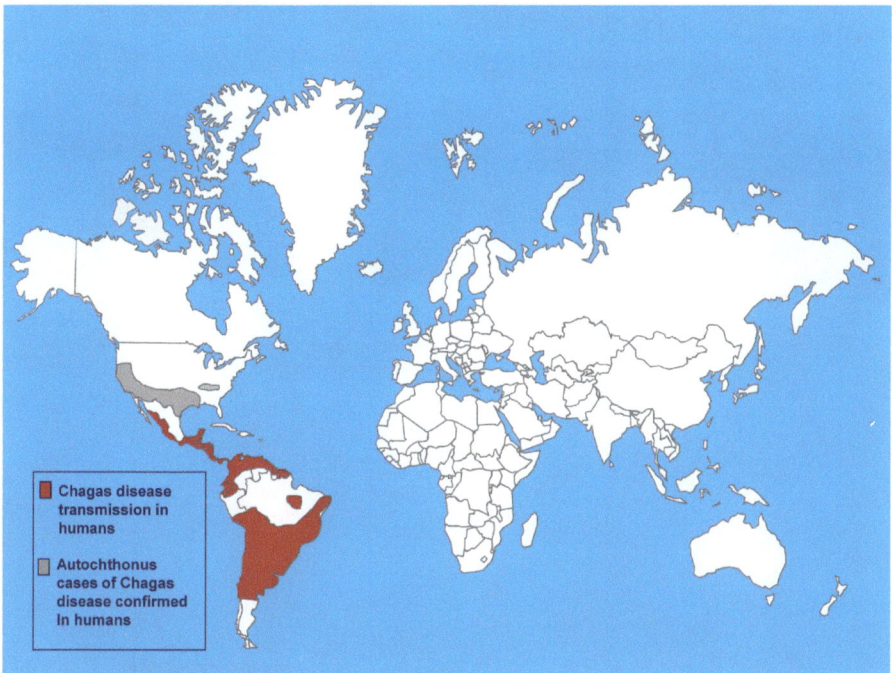

Chagas disease cases worldwide. Source: Adapted from the World Health Organization

The infamous tsetse fly transmits another parasite that causes sleeping sickness (African trypanosomiasis), with an estimated 65 million people at risk in 36 sub-Saharan African countries. Throughout Latin America and the southern United States, kissing bugs (Hemiptera) carry the parasite that causes Chagas disease, placing roughly 100 million people at risk with ~16 million infections. Similarly, black flies transmit the parasitic worm that causes onchocerciasis, or river blindness, with 16 million people infected in Africa, and Central and South America. Finally, the historically dreadful flea-borne disease bubonic plague continues to be active around the world.

Appendix 6 provides an extensive worldwide listing of known dangerous invertebrates.

Invertebrate threats

Invertebrate threats encompass two broad categories: point source threats and psychological threats. Point source threats cause physical injury, distress, or death within a brief time. A wasp sting, deadly disease such as mosquito-transmitted malaria, and allergic reactions are examples of invertebrate point source threats. Psychological threats do not kill or directly threaten health, but they may make someone uncomfortable, and impair normal activity and behaviour. Both types

of threats can disrupt, or even end, your travel plans. The following are the various responses that may result from contact with invertebrates.

Physical pain

A widely diverse array of invertebrates can inflict bites, piercings, pinches and stings that can cause various degrees of suffering. Symptoms range from mild annoyance to incapacitation. Although generally not lethal, such physical trauma may impede

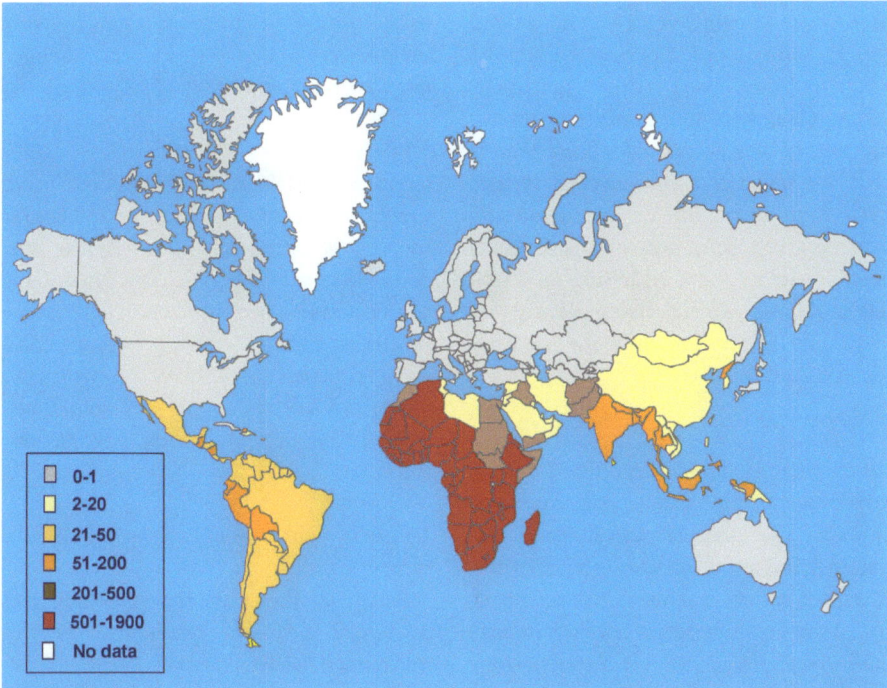

Legend:
- 0-1
- 2-20
- 21-50
- 51-200
- 201-500
- 501-1900
- No data

Vector-borne disease deaths per million people. Source: Modified from USA Centers for Disease Control and Prevention

normal activity, or even prove psychologically disturbing.

Disease

The most serious invertebrate-related risk is arthropod-borne or vector-borne disease transmission. The World Health Organization (WHO) estimates 10 million vector-borne disease cases occur annually worldwide, many of which are fatal. Some of the most significant vector-borne diseases include malaria, yellow fever, dengue, chikungunya, Japanese encephalitis, filariasis, American trypanosomiasis (Chagas disease), Leishmaniasis, tick-borne encephalitis and schistosomiasis. There are many others.

Envenomation

Invertebrates can introduce a toxin (venom) directly onto or into a person via bites or stings, and the impact is often immediate. Venoms may be either neurotoxic and/or necrotic, or have properties of both. Neurotoxic venoms affect the nervous system, while necrotic venoms destroy blood and tissue. Impacts range from localised envenomation site reactions including mild irritation and/or limited tissue damage to body-wide (systemic) reactions such as organ failure and even death.

Myiasis

Larvae (maggots) of certain flies (Diptera) can invade living tissue to consume flesh or body fluids. Myiasis may cause substantial psychological distress in victims. Although certain strains of carrion-eating fly larvae are sometimes used to remove dead wound tissue (maggot therapy), some closely related species may also damage living, healthy tissue. The invasion of tissues by flies may be benign or even asymptomatic, but often it is destructive to the patient. Types of human myiasis include urogenital, gastrointestinal, ocular, auricular and cutaneous. Cutaneous myiasis is the most common form and can be quite dramatic in appearance. In the case of creeping myiasis (creeping eruption), the progress of the maggots can be tracked as they burrow under the skin. Traumatic myiasis involves wounds (lesions), and furuncular myiasis is the development a boil-like lesion.

Allergic reactions

These reactions may be localised (wheals, swelling) or they may be systemic (anaphylactic shock). Severity varies greatly, and may include death. Most allergic reactions occur primarily through contact with invertebrate venom, saliva or certain body parts such as setae.

Urtication

Certain invertebrate body parts including the setae of certain moth larvae and nematocysts (stinging cells) of jellyfish and corals have toxins that cause a physiological response when contacted. Urtication reactions include painful burning and itchy skin eruptions or hives (i.e. urticaria). These can be debilitating, and sometimes develop into systemic shock, but they are rarely fatal.

Delusory parasitosis and entomophobia

These phenomena are true psychological disorders. Delusory parasitosis is the unfounded belief that parasites, usually insects, are living on or in the body. Reactions and responses are often agitated, distressing and intensely emotional. This condition, although very rare, can be incapacitating, and may require professional mental health care. Entomophobia, by comparison, occurs only in the presence of certain insects. It is simply an overwhelming, irrational fear of insects and spiders or their relatives, following a previous unpleasant encounter. For example, some people may develop an irrational fear of bees after being stung. Both psychoses are often cumulative and become increasingly exaggerated in the absence of professional care. Increasing negative experiences may promote increased negative health and welfare. This cumulative effect is a function of stress, fatigue, living conditions, pest diversity and density, and the ability to escape or avoid contact with invertebrates. Thus, under certain conditions (high pest densities, low disease incidence) these psychological disorders can become a greater threat to wellbeing than the actual disease.

Personal protection measures

Whether your adventure involves local or foreign travel, outdoor adventures or working around home, it is important to learn about the hazards that may be faced and ways to properly protect yourself. Some protection measures are common sense, such as only using approved food and water sources, not swimming in contaminated water, avoiding close contact with animals (wild, domestic or stray), reporting animal bites and practising good personal hygiene. There also are serious invertebrate hazards that may go unseen, including night-feeding mosquitoes or attached ticks that require proper personal protection methods and equipment.

There are six general classes of protection measures: chemoprophylaxis, vaccination (immunisation), avoidance, barriers, repellents and personal use pesticides.

Chemoprophylaxis

This is medication taken before exposure to vectors of a disease. The medicine circulates through your body to kill the introduced pathogen. They do not change the immune system so they must be taken before, during and after travel to disease risk areas (see 'Education' section). Selecting the proper chemoprophylactic agent is based on an individual health risk assessment and a disease risk assessment for a specific area. Some types of chemoprophylaxis must be taken several days before any potential exposure in order to be effective. Planning is therefore crucial in order to maximise benefit. Adventure plans that involve areas with high disease risk should include seeing a health-care provider or travel medicine clinic no less than 6 weeks before travel commences. Be sure you bring enough medicine to last the entire time estimated for travel. Chemoprophylaxis should be obtained only from a reputable pharmacist or physician, because there are many ineffective counterfeits sold in some foreign markets. In addition, a follow-up assessment after travel is completed and additional treatment may be required. Malaria is a good example of such a scenario. Depending on which malaria parasite causes an infection, chemoprophylaxis does not necessarily kill it. Malarial prophylactic drugs including mefloquine, primaquine, chloroquine and doxycycline control the symptoms while travelling, but the patient may need additional curative treatment under direct practitioner supervision upon returning home. Otherwise, the infection can flare up days to months after the chemoprophylactic medication is no longer taken. Thus, it is a common occurrence for travellers to return home without any symptoms, not knowing the infection remains, or that the infection will soon demonstrate symptoms. So scheduling a follow-up appointment with a medical provider is highly recommended before even travelling to a high-risk area. An additional consideration is that chemoprophylactic medications are not always 100% effective, and some pathogens have developed resistance to certain medications. A medical practitioner should be able to provide a treatment regimen based on the most current available information for the region of interest.

Vaccination (immunisation)

Unlike chemoprophylaxis, vaccination changes your immune system by introducing non-pathogenic live or killed disease organisms in very small doses before exposure allowing your body's natural defences to resist the disease. However, this process takes time. No vaccine provides immediate protection and some require periodic boosters to remain effective. For example, the yellow fever vaccine takes 10–11 days after injection to provide maximum protection. However, once it becomes effective, the yellow fever immunisation lasts up to 18 years and a booster is given every 10 years for assurance. Your health-care provider or travel medicine clinic should be visited 6 weeks before you go to areas with a disease risk, and schedule a follow up assessment. Similar to chemoprophylaxis, no vaccine is 100% effective.

Avoidance

There are many common sense ways to avoid venomous and medically important invertebrates including:

- **Avoid disease vector and pest dwelling places.** Before travel, learn about places where you might encounter dangerous invertebrates and how to avoid them. For example, kissing bugs that carry Chagas disease often inhabit thatched roofed and adobe walled structures throughout their range. Congo floor maggots occasionally inhabit the mud

floors of dwellings. Tsetse flies are highly attracted to motion, including moving vehicles.

- **Avoid vector and pest breeding places.** Similarly, try to avoid areas where mosquitoes breed such as swamps, or, in the range of black flies, avoid flowing streams.

- **Avoid terrain features that attract or harbour pests and vectors.** Many vectors and pests, such as ticks, fleas and mosquitoes, can be found along transitional vegetation including trails, brushy areas and meadows where warm-blooded animals tend to dwell.

- **Avoid pest and vector peak activity periods.** It is important to understand the activity behaviours of the vectors and pests in the area where you are going. For example, some *Aedes* mosquitoes bite during daylight while other species feed at dusk and dawn. Avoiding peak activity periods is important for minimising exposure risks.

- **Avoid exposing unprotected skin.** The best way to avoid skin exposure to invertebrates is to use personal protection approaches.

- **Avoid bringing them home.** Medically important vectors/pests can travel home with you so it is important to follow a few simple rules to avoid this, particularly after travel to high-risk areas. Some vectors/pests can live for up to a year off their host. For this reason, we recommend unpacking outside (e.g. garage, carport) or away from living areas such as bedrooms (i.e. use the laundry room, utility room, etc.). It is always a good idea to dry clean or wash all travel clothing (clean or dirty) in hot water with detergent as soon as possible upon return and carefully inspect baggage for signs of invertebrate hitchhikers. If these actions cannot be accomplished in a timely fashion, we recommend storing luggage and other travel items in plastic trash bags until you are certain all potential hitchhikers are dead (depending on the potential travel-risks you faced). Chemically treating suspect baggage with pesticide products labelled for that purpose may be necessary.

Barriers

The concept of barriers is simple. It involves placing something physical between you and the vector and/or pest as a means of eliminating contact. Examples include:

- **Oils** (e.g. baby oil, skin softener). These are short-lived physical barriers that typically do not repel or kill and require frequent reapplication, especially during hot weather. For example, the widely touted Avon moisturiser Skin-So-Soft[R] bath oil protects against mosquitoes less than 30 minutes and does not protect against ticks at all. It, like all oils, does not repel. Oils clog the arthropod's sensory organs disrupting attempts to bite. There are situations when oil is a good option such as when a registered repellent approved for use on children under 2 months old (see Appendix 2) does not exist. In such cases, a wide variety of oil types will work, including vegetable or mineral, although you should be alert for possible sensitivities to particular types of oil.
- **Screens.** Window screens, door screens and tent screens are all effective barriers against invading invertebrates. You can also improvise and use bed-netting material over an open window in the absence of other screens to help keep pests/vectors away. If possible, use a residual pesticide such as permethrin to treat screens since some very small vector/pests can go through normal-size screen.
- **Shelters.** Use fixed structures with screened and sealed windows and doors to the extent possible. Temporary structures such as tents and hammocks also are effective, but they should have sealed floors and a zippered fly. The best tent structures typically have two doors and a vestibule designed to make it difficult to open both doors simultaneously. Regardless, to test the integrity of a tent against invading invertebrates, you should conduct a 'light leak' test to find entry points. During daylight hours: douse all internal light sources; close or cover windows, flaps, and so on; let your eyes adjust; then find any pinholes or cracks. Particularly look around doors and windows, along floor, under beds/cots and ceiling corners. Use duct tape or other suitable material to close gaps.
- **Bed net.** Bed nets can be highly effective for protecting against invertebrates. Always use a bed net in high-risk areas. Wide varieties of bed nets are commercially available at a reasonable cost. Most are easily transported in luggage, and some are designed similarly to tents or pop-up designs. The keys for obtaining an effective bed net relate to mesh size, tucking loose edges underneath the bed, and treating the net with permethrin. Because dangerous invertebrates can crawl under suspended

netting, or even netting touching the floor, always tuck the net under bedding or mattress so bodyweight forms a seal. Ideally, a fine mesh ('tropical mesh') bed net should be used because some small vectors and pests (sandflies, gnats, mites, etc.) can pass through larger mesh netting. In addition, since bare skin may touch the net during sleep, permethrin treatment is necessary to help prevent biting through the net. The best bed nets are completely enclosed, sealed (i.e. zippered), permethrin treated, and fine meshed, such as the 'pop-up' bed nets or tropical hammocks. Even then, before entering the bed net, check around, and under it for the presence of invertebrates. In addition, an approved 'knock down' (readily dissipating) aerosol insecticide should be used once you are safely enclosed inside of the bed net to kill any unseen organisms. Always leave a bed net tucked in and sealed so pests do not enter while you are away.

- **Head net.** Head nets can work well, and a wide variety are commercially available,

Head net. Source: Grace E. Bowles

but, for them to function properly, you must wear a hat. That said, they are prone to snagging and can block vision, and they can be hot to wear. The best head nets are fine-meshed with a chest trap to hold the net against the body so organisms cannot crawl under. Permethrin treatment prevents bites

Mosquito bed net. Source: Tjeerd Wiersma/ Wikipedia, CC BY 2.0. Available from https://en. wikipedia.org/wiki/Mosquito_net#/media/ File:Mosquito_Netting.jpg

Commercial head net. Source: David E. Bowles

through the netting. Many head nets are designed so that they can be worn while sleeping. Sometimes head nets are essential, such as when fishing or hunting around black fly and mosquito swarms, or tabanid flies.

- **Protective clothing.** Many harmful invertebrates are attracted to body odours (carbon dioxide, octenol, lactic acid, etc.) that settle to the ground. For example, once attracted, ticks may crawl up the body seeking a point where they can attach and feed. This particular behaviour can be restricted by tucking pants into socks or footwear (boots are best). In vector risk areas, you should resist the temptation to wear shorts, short-sleeved shirts, sandals or open-toed footwear. Properly worn clothing can provide an excellent barrier to invertebrates. For example, a loose fitting, long-sleeve shirt or jacket, long pants, hat and boots work best for preventing biting pests. Light-, neutral-coloured and at least medium weight fabric is best, including socks. In addition, light coloured clothing makes it easier to see and remove certain pests, such as ticks.

Many invertebrates can bite through thin cloth (e.g. mosquitoes, tsetse flies, tabanids, fleas) and many are attracted to bright colours (blue, yellow, red), very dark colours, floral patterns, and metallic fabric (e.g. tsetse flies, black flies, mosquitoes, tabanids). Because some vectors/pests can bite or sting through even heavy fabric, use of permethrin treated clothing is recommended. Treating clothing with permethrin can be done as do-it-yourself, but care must be taken to

Author James Swaby shown in Kenya with properly worn protective clothing to protect against arthropod attacks. Source: Department of Defense

follow label directions on both the clothing and permethrin, and even greater care must be taken to ensure the pesticide is evenly applied. A wide variety of commercially available clothing that is factory treated with permethrin or similar compounds can be very useful for preventing attacks by biting and stinging insects. Such clothing includes shirts, pants, socks, hats and jackets.

Finally, a simple and practical means of preventing arthropod attacks is to use wide tape (duct or similar) wrapped around ankles and even thighs when certain vector/pest populations (e.g. fire ants, ticks, fleas, mites) are extremely high. Wrap the leg or ankle

A commercially available permethrin-treated sleeping sack. Source: David E. Bowles

Commercially available insect resistant outer pants.
Source: David E. Bowles

once with the sticky side towards the body then around again twisted, so the sticky side faces out. The pests become

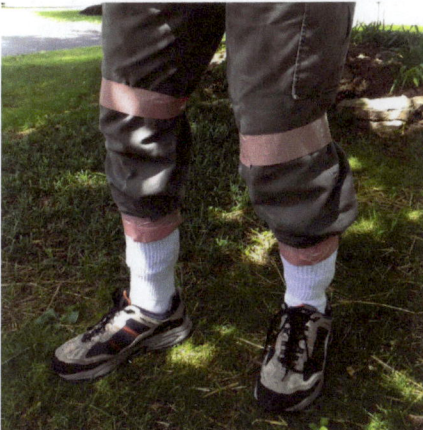

Sticky tape method of preventing tick and chigger bites. Note the tape is wrapped around the ankle where the pants are tucked into the socks, and again around the thigh above the knee. Source: Grace E. Bowles

stuck to the tape, which can be replaced every few hours for maximum effectiveness. Although using tape in this fashion may make you look ridiculous, the results are generally highly effective. Wide varieties of duct tape colours are now available, including camouflage.

Repellents

Repellents are chemicals applied to skin, clothing or other surfaces to make you less attractive to arthropods. The perfect repellent would: provide 100% protection against all biting arthropods for several hours under different environmental conditions; and be non-toxic, non-irritating, harmless to clothing, cosmetically acceptable (feel good, smell good), easy to apply and inexpensive. There currently is no perfect repellent. Despite advertised claims, none is 100% effective alone or in combination with other methods, no matter how much you use. Finally, just like chemoprophylaxis and vaccinations, you cannot just trust your health to repellents alone. You must also use other personal protection methods to maximise protection against arthropods.

Choosing a repellent

This is a very dynamic area, with new products available on the commercial market, frequently changing test data, warnings and recommendations. Choosing a safe and effective repellent can seem like an onerous task. The only way to make a fully informed decision is to check the informational sources listed under 'Education', all of which rely on scientific, peer-reviewed data. To aid the reader in this effort, Appendix 1 provides United States

Environmental Protection Agency (EPA) registered repellent efficacy and use assessments. Appendix 2 provides safety data and safety assessments. Appendix 3 summarises commercially available EPA registered repellents. The EPA reviews and approves any asserted efficacy, and use and safety claims. Most of these repellents or versions of them are sold in many markets worldwide. Make sure you use products with an effective concentration, so always read and follow the EPA approved product label directions. Labels may use the chemical name such as N, N-diethyl-m-toluamide, or an abbreviation such as DEET. Also, make sure the label lists the pest you want to repel and then adhere to restrictions on age, pregnancy and breast-feeding (see United States Centers for Disease Control and Prevention (CDC) and EPA). Be very cautious about 'personal experience' recommendations, manufacturer endorsements for products that have not been thoroughly vetted, and popular and social media sources.

Repellent recommendations

Many formulations of repellents are available with differing active ingredients, and concentrations. You should select a repellent based on your anticipated adventure and potential exposure risks. Many 'real-world' factors impact how long repellency lasts, including perspiration, humidity, ambient temperature, body temperature, surface abrasion due to rubbing or washing, wind, rainfall, wading or swimming. In other words, the repellent chosen for a rigorous hike may be different from one selected for a relaxed outdoor social gathering. The keys for selecting the most appropriate repellent include the effective concentration of active ingredient, estimated length of protection, advantages, disadvantages, limitations, restrictions and safety. High concentrations of active ingredient do not make the repellent the best choice and concentrations greater than 50% do not necessarily perform better than lower concentrations. The best repellent choices are 'extended duration' products that microencapsulate the repellent to control the release rate while resisting sweat, moisture and environmental effects. These repellents also reduce skin absorption, and you need to apply much less product and far less frequently.

Only a small amount of lotion type repellent is needed for effective protection. The approximate amount necessary to treat hands, face, neck and ears (the only areas exposed when you properly wear protective clothing) should cover the tip of the index finger. The two EPA registered 'extended duration' formulations are the 3M Ultrathon (polymer-based microencapsulated 33.4% DEET) and Sawyer's Controlled Release (sub-micron protein

Approximate amount of lotion type DEET product needed to treat hands, face, neck and ears. Source: James A. Swaby

Mosquito coil in use. Source: Jo Naylor/Flickr. CC BY 2.0. Available from https://www.flickr.com/photos/pandora_6666/9500345872/

microencapsulated 20% DEET). Products containing picaridin have been shown to have efficacy comparable to that of DEET. These are the most advanced, reliable products under 'real-world' conditions.

Commercially available insecticidal coils containing pyrethroids or other compounds including citronella and sandalwood extract have been shown to have some efficacy against flying pests such as mosquitoes. These coils work by burning and releasing the pesticide in smoke, which then permeates the air space. Such coils can be purchased in many areas of the world. Scientific studies have shown a broad range of effectiveness among the various commercial brands, with some working reasonably well but others only poorly so.

Other commercially available repellent products for protection from biting mosquitoes are 'clip-on' devices and wristbands. Clip-ons typically contain pyrethroid pesticides and are designed to attach to clothing, belts, and so on. Some are advertised to have long duration up to

12 hours, may have a battery powered fan to help distribute the repellent, and some are refillable. Clip-ons offer the advantage of not having to spray or apply repellent directly to the skin. A disadvantage is that people and pets can inhale the dispersed pesticide. When mosquito populations are low, or in low-risk areas such as backyards, these devices may work reasonably well. A related device is the insect repellent wristband. These products may contain pyrethroid repellents or natural products such as peppermint oil. While these wristbands may provide some repellency effect in the area near the wristband, they are generally considered ineffective. In areas with large mosquito populations, or where disease transmission may occur, we do not recommend using clip-ons or wristband products

Some products marketed as electronic or sonic insect repellents are touted as being effective for repelling mosquitoes and other biting insects. These products are essentially ineffective and we do not recommend their use at all.

Repellent use precautions

The following recommendations will maximise personal protection from biting insects, while emphasising personal safety:

1. Be cautious the first time you use a repellent. Everyone is different and you may be allergic to ingredients so apply a small amount and watch (15–30 minutes) for any reaction. Immediately discontinue repellent use if you have any adverse reaction, wash treated skin, follow the label first aid directions and seek medical attention (bring along the repellent that caused the observed reaction).

2. Use sparingly, do not over apply, a small amount is usually effective. Follow the label re-application times, and avoid application to mouth, eyes, cuts, wounds, sunburns or irritated skin.

3. Apply to hands first and then rub on face.

4. Apply only to exposed skin. Do not apply to clothing covered skin (see clothing treatment section).

5. Children should not apply repellent; adults should apply it for them. For children, avoid applying it to the eyes, mouth and palms.

6. Use soap and water to remove repellent once you are no longer outdoors, particularly after repeated daily or consecutive daily use.

7. To use insect repellent with sunscreen, apply the sunscreen first then the repellent so the sunscreen does not block the repellent or increase repellent absorption. Applying DEET/picaridin first can also reduce the sunscreen's SPF factor. Using products containing both sunscreen and repellent can be disadvantageous because the sunscreen may require re-application at a greater frequency than the repellent. In the combination products, this may result in over application of the repellent.

8. Similarly, for hunters choosing to wear camouflage face paint, apply the camouflage paint first and then the repellent.

9. Avoid repellent wristbands and clip-on repellents. They are not as effective as repellents applied to the skin or clothing, or offer little measurable protection.

10. Avoid insecticidal fans (containing metofluthrin, allethrin), outdoor foggers, repellent candles, and so on.

They do not provide lasting protection while posing an inhalation hazard.

11. Avoid electronic or 'sonic' pest repelling devices. They do not work at all for the intended purpose.

Concerns about DEET

You may have concerns about using DEET, especially for children, pregnant or nursing women. The peer-reviewed, scientific and medical literature describes a few allergic, toxic and neurological reactions to DEET. In addition, some individuals are more sensitive to chemicals than others are and repeated applications have occasionally produced tingling, mild irritation or contact dermatitis. Rarely, toxic encephalopathic reactions to DEET have been documented, but these cases were a result of exposure to high concentrations DEET due to over application and oral consumption. A 1998 EPA review of reported seizure cases concluded that adverse reactions rate appeared very low, about one per 100 million people.

All said, after over 50 years of use and an estimated 8 billion applications, DEET is the most exhaustively studied repellent available and has a remarkable safety record. Since 1960, the medical literature reports fewer than 50 adverse reactions to DEET among some 912 million applications. Roughly 25%, or 24 million people, were children with only 14 documented adverse reactions. Thus, assuming that potential DEET toxicity cases are grossly under reported, the odds are many millions to one against any toxic DEET reactions among children, much less adults. Ten to 30% concentrations of DEET are considered safe and effective for pregnant or breastfeeding women and children over 2 months old (Appendix

1–3). Regardless, there is no 100% safe substance: only safe ways to use them. You should always read and follow the product label to ensure maximum safety.

'Natural' botanicals, dietary supplements and sulfur

These repellents include castor, cedar, citronella, clove, geraniol, lemongrass, patchouli, peppermint, rosemary and soybean oil. Botanicals are exempt from EPA registration because they are classified as 'minimum risk' pesticides, which means they are not required to conduct safety or efficacy testing. Moreover, 'minimum risk' does not mean they are safe for use and there is no assurance they actually work for the intended purpose. There simply is not much data to confirm or contradict the advertised claims. There are a couple of EPA registered botanical repellents on the market (Appendix 1–3). However, these botanical-based products often contain higher product concentrations than related personal care products, and they can contain known human allergens (see Appendix 4). Some are irritants, carcinogenic (e.g. geraniol), or even acutely toxic at high concentrations. Compared with EPA registered repellents, supplements such as garlic, onions, radishes, vitamin B do not effectively repel arthropods. A similar lack of proven safety and effectiveness goes for sulfur-based tick and chigger repellent products.

Clothing treatment

You can apply any repellent to clothing, but most are not designed for this purpose. Permethrin is the only EPA registered repellent specifically formulated for clothing application. Applying other types of repellents not designed for this purpose is not nearly as effective. Permethrin is both a repellent and an insecticide, but it is intended only for fabric treatment including clothing, outerwear, bed nets and tents. Do not apply permethrin to your exposed skin. Permethrin-treated clothing, properly worn with an extended duration repellent on your skin, can provide you maximum protection from biting arthropods. Permethrin will not bind to Nomex® or Gortex® fabric.

Commercially available permethrin clothing treatment kits (e.g. Sawyer's) contain 0.5% permethrin with an effective duration of around 5 weeks. Wearing permethrin-treated clothing is recommended for people who spend a lot of time outdoors or frequently encounter biting pests. Several companies sell permethrin treated clothing, but the effectiveness of these products varies greatly depending on the method used to treat the fabric.

Personal use pesticides

Use commercially available, EPA (or equivalent) approved pyrethroid aerosol space sprays to control arthropods inside of bed nets, tents, closed rooms, vehicles, and so on. Examples are permethrin, d-phenothrin, resmethrin and pyrethrum. Always read the product label and use only as directed.

Education

'Know before you go.' In other words, before travel is initiated, you should ideally have a basic understanding of what vectors/pests and diseases might be encountered during travels (both domestic and foreign). Specifically, a health risk assessment should be prepared and any necessary vaccines/chemoprophylaxis should be obtained before leaving for the trip. Consult your personal physician, a travel health clinic or other medical professionals, preferably

6 weeks before you travel to allow adequate time for prophylactics and vaccines to work. Also, read and understand current travel destination advisories and recommendations (e.g. the Australian Government Smart Traveller website, CDC, PAHO, WHO, US State Department Traveller's Alerts, or other reliable sources). We recommend all travellers carry a personal health summary, including vaccination record, and wear any necessary Medic-Alert tags. Emergency health personnel need to know your allergies, medications, medical history and physical condition so they can determine appropriate and accurate treatment, if needed. Appendix 5 shows an example of a Personal Health Summary. Finally, many arthropod-borne diseases (Chagas, Zika, dengue, trypanosomiasis, malaria, yellow fever, encephalitis, etc.) can be acquired through blood transfusion, organ transplant or trans-placental (fetal) pathways. If emergency medical care is obtained in high-risk areas, a follow-up appointment with a personal physician should be conducted to assess potential risks of those treatments. Pregnant women must be particularly careful travelling to high-risk areas. To minimise risk, a physician's travel guidance and recommendations should be followed.

Things to avoid

Social and popular media are full of 'folk remedies' to prevent arthropod bites and disease transmission. Avoid these products and methods because they generally are not based on peer-reviewed, rigorous scientific test data, and they are invariably ineffective and unsafe. For example:

1. Wearing flea collars. Wearing these products around the neck or legs is unsafe. They are designed for hairy pets and are not intended for people. Pesticide is absorbed through the skin, and the wearer may suffer significant dermatological distress.

2. Taking mega doses of vitamins. There is no evidence they repel or make you less attractive (to pests), and doing this may actually result in poisoning.

3. Eating garlic, onions or radishes. While the compounds in these plant parts may repel a few mosquitoes, they generally do not work.

4. Drinking vinegar to repel mosquitoes. This method does not work and it can be toxic when consumed in large amounts (see LD_{50} footnote in Appendix 2).

5. Using baby oil alone or mixed with perfumes as a repellent. Oil is a barrier, not a repellent, and sweet or flowery smelling fragrances actually tend to attract arthropods, not repel them.

6. Eating match heads to repel chiggers and ticks. The concept here is that the odour from the sulfur secretes through the skin and repels the pests. For this method to work, you would have to eat enough matches to smell like sulfur, which could result in illness or death.

Sponges
(Phylum Porifera)

About a dozen marine sponge species in eight families are considered toxic and can cause contact dermatitis in people. They primarily inhabit warm Caribbean waters but also the North Atlantic Ocean off North America and Europe, and the Pacific Ocean off California, Mexico, and Australia. However, potentially dangerous sponges may exist elsewhere.

Accidental contact and handling can result in serious consequences for the victim. For example, the fire sponge (*Tedania ignis*) is commonly implicated in negative human reactions. The typical reaction is an almost immediate skin irritation, redness and contact dermatitis (similar to poison ivy) where the sponge contacted the skin. Finger joint stiffness (if handled) and localised swelling may follow. Blisters often develop within a few hours.

Diagnostic testing is seldom necessary but involves placing a small part of the suspect sponge on an unaffected body part to observe any reaction. Antiseptic lotions or dilute acetic acid (vinegar) helps ease the itching and burning. Blisters may require antibiotic ointment.

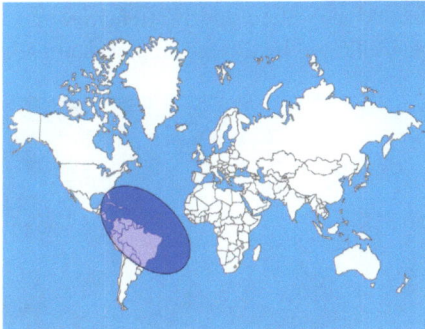

Fire sponge (*Tedania* sp.). Source: Ed Bierman/Flickr. CC BY 2.0. Available from https://www.flickr.com/photos/edbierman/9058286938/

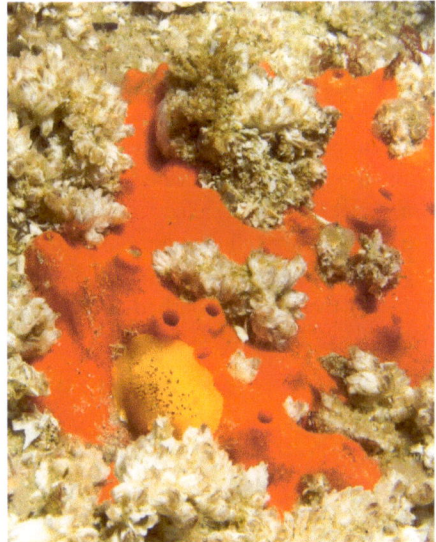

Distribution of fire sponge (*Tedania ignis*).

Coelenterates
(Phylum Cnidaria: jellyfish, sea anemones, sea fans, corals)

Over 9000 coelenterate species inhabit mostly marine systems worldwide including free-swimming true jellyfish medusae and fixed forms such as corals, sea ferns and sea anemones. A few medically unimportant species are freshwater forms. You may encounter beaches displaying cnidarian warning signs because many can produce painful and dangerous stings. Some jellyfish are extremely dangerous and are capable of killing you.

Coelenterates are among the most primitive carnivorous animals. Despite their simple and delicate body forms, they have developed sophisticated envenomation mechanisms called nematocysts. Nematocysts are cellular capsules containing a coiled, barbed tube (see diagram). When activated, this barbed tube penetrates you, injecting venom. A single cnidarian can have millions of nematocysts.

Jellyfish
(Subphylum Medusozoa)

There are over 2000 described jellyfish species but only ~70 can cause serious injury. The common and widely distributed moon jellyfish (*Aurelia aurita*) is generally considered harmless. However, it occasionally stings people, particularly in the Gulf of Mexico. Such encounters are more likely near coastal areas where populations are greater. Stings cause instant localised pain lasting up to 30 minutes. Hives and inflammation may appear around the wound shortly after envenomation followed by ulceration. Residual pain

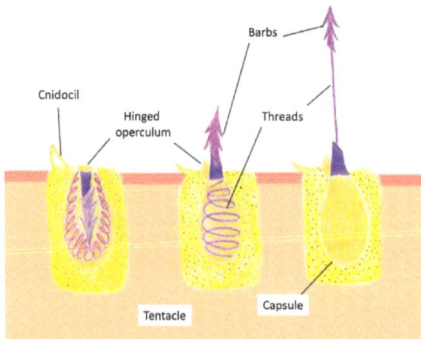

Diagram of a coelenterate nematocyst showing the discharge mechanism. Source: Redrawn from Spaully/Wikipedia. Available from https://en.wikipedia.org/wiki/Cnidocyte#/media/File:Nematocyst_discharge.png

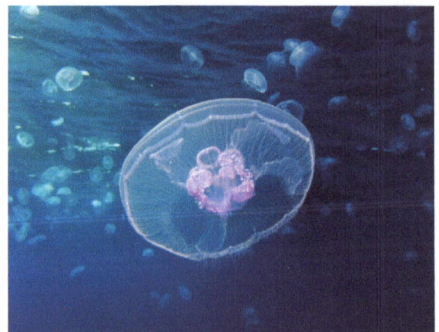

Moon jellyfish (*Aurelia aurita*) at Gota Sagher (Red Sea, Egypt). Source: Alexander Vasenin/Wikipedia, CC BY-SA 3.0.0. Available from https://en.wikipedia.org/wiki/Aurelia_aurita#/media/File:Moon_jellyfish_at_Gota_Sagher.jpg

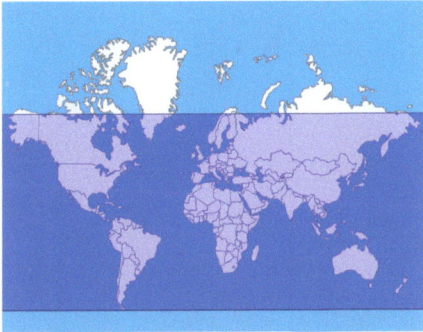

Distribution of moon jellyfish (*Aurelia aurita*).

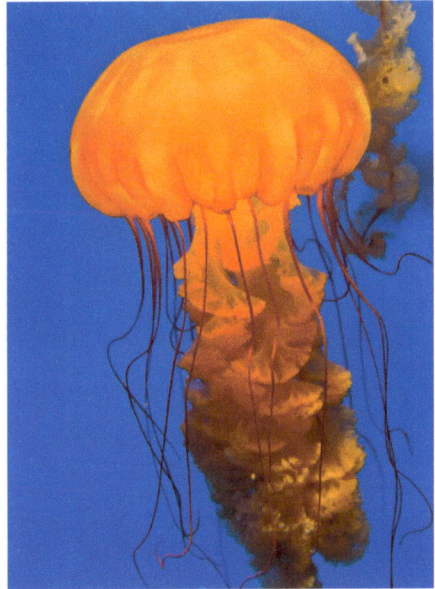

East coast sea nettle (*Chrysaora quinquecirrha*). Specimen appears orange due to lighting. Source: Antoine Taveneaux/Wikipedia, CC BY-SA 3.0.0. Available from https://en.wikipedia.org/wiki/Chrysaora_quinquecirrha#/media/File:Chrysaora_quinquecirrha.jpg

may last several days. Encrusted lesions become obvious in a few days and post-inflammatory, darkly pigmented skin may last up to 2 weeks.

Another common and widely distributed jellyfish, the East coast sea nettle (*Chrysaora quinquecirrha*) occurs in the Atlantic Ocean from Cape Cod south along the United States eastern coast into the Caribbean and Gulf of Mexico, and in portions of the Indo-Pacific Ocean. It was also introduced to the Black Sea in Europe. It is commonly found near the confluence of coastal tributaries and bays with 10 to 20 ppt salinities. It is generally white in colour, but can have prominent maroon-coloured markings in some areas. East coast sea nettle stings, while painful, are not considered deadly.

A related species, the West coast sea nettle (*Chrysaora fuscescens*) has an equally painful sting. It ranges from British Columbia to Mexico although it is most commonly found in the northern and eastern Pacific Ocean. It often forms massive swarms with nearshore aggregations most common during autumn (fall) and winter months. West coast sea nettles have a distinctive golden-brown bell (up to

30 cm diameter) with whitish oral arms. Thin maroon tentacles may trail behind several metres.

Various, potentially dangerous, box jellyfish occur in warmer oceans worldwide,

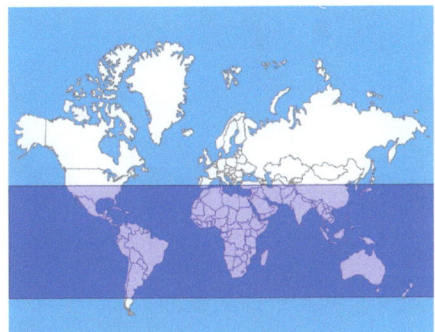

Distribution of East coast sea nettle (*Chrysaora quinquecirrha*).

West coast Sea nettle (*Chrysaora fuscescens*).
Source: Ed Bierman/Wikipedia, CC BY 2.0. Available from https://en.wikipedia.org/wiki/Chrysaora_fuscescens#/media/File:Sea_nettle_(Chrysaora_fuscescens)_2.jpg

Box jellyfish (*Chironex fleckeri*). Source: Guido Gautsch/Wikipedia, CC BY-SA2.0. Available from https://en.wikipedia.org/wiki/Chironex_fleckeri#/media/File:Avispa_marina_cropped.png

but the box jellyfish or sea wasp (*Chironex fleckeri*) is considered among the most dangerous marine animals in the world (if not the most dangerous). It has been implicated in over 70 deaths throughout its Indo-Pacific Ocean distribution (primarily around Australia). Some consider it the most venomous creature on the planet.

Children are particularly vulnerable. Most stings occur on humid days when the water is calm. November to May is generally considered unsafe for swimming in nearshore tropical Pacific waters due to

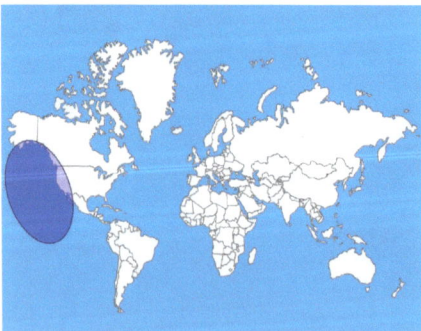

Distribution of West coast Sea nettle (*Chrysaora fuscescens*).

this jellyfish in addition to the highly dangerous Irukandji (see later). *Chironex* prefer calm waters within 2 km (1.2 miles) of the shore and tend to congregate near estuaries. They are rarely found in open waters. On windy days, they drift below the choppy surface to calmer, deeper waters where unsuspecting swimmers encounter them. Despite claims to the contrary, sting netting used as a deterrent is not 100% effective because stings occur each year within so called 'safe swimming' enclosures.

Chironex stings are severely painful, peaking in 15 minutes but persisting up to 12 hours. Immediate rapid heartbeat and high blood pressure are superseded by an unusually slow heartbeat with inadequate blood flow causing shock, other heart anomalies and pulmonary oedema. Also, neuromuscular paralysis may lead to respiratory arrest. Within minutes, loss of con-

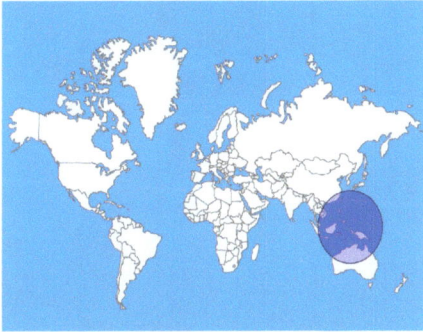

Distribution of box jellyfish (*Chironex fleckeri*).

sciousness and death may rapidly ensue. The sting pattern is characteristically visible as a frosty appearing whiplash, beaded, or a ladder pattern of red, purple or brown skin lesions. Among survivors, these develop over several days into ulcers and widespread tissue necrosis. They slowly heal over several months but often with significant residual pigmentation and scarring.

First aid for box jellyfish and Irukandji must commence as soon as possible. Treatment is mainly supportive. The traditional remedy of liberally pouring vinegar over the wound to inactivate the nematocysts may have little value and provides minimal relief at best. Several other substances such

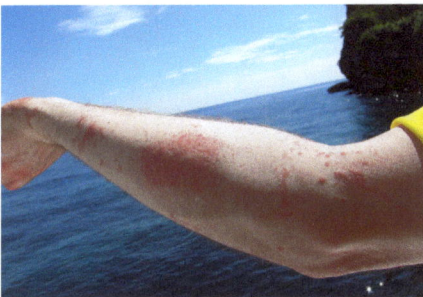

Jellyfish sting. Source: Thomas Quine/Flickr. CC BY 2.0. Available from https://www.flickr.com/photos/quinet/6052293114/

as freshwater, alcohol, tea, urine, cola drinks and aluminium sulfate (Stingose®), also are not effective and may activate undischarged nematocysts. Similarly, there is no evidence that compression bandages and limb immobilisation modify the subsequent clinical course. Cautiously, manually remove adhering tentacles to prevent further stinging to victim and yourself. Basic life support measures, including cardiopulmonary resuscitation (CPR) and hospitalisation may be required. Pain suppressants are often administered during transport to hospital. Treat irregular heartbeat with appropriate agents and pain control generally requires large intravenous narcotic analgesic doses. Several treatments with morphine or opiates may be required before the pain subsides. In the most serious cases, an extremely dangerous sympathomimetic syndrome may develop with a broad suite of characteristic symptoms including delusions, paranoia, rapid or slow heartbeat, irregular heartbeat, high or low blood pressure, high fever, sweating, bristling of hairs, dilated pupils, overactive physiological responses and seizures. This condition requires immediate emergency medical care involving advanced life support measures such as respiratory ventilation with continuous positive pressure airway or tracheal intubation. Treat skin and tissue lesions conventionally, but they may eventually require surgical debridement and grafting.

Carukia barnesi and *Malo kingi*, commonly known as Irukandji (an Aboriginal tribal name) or Kingslayer in Australia are small species of box jellyfish that are widely distributed in open and coastal waters of the southern Pacific. A few other species of *Malo* are distributed in the greater

Indo-Pacific Ocean and they also may be capable of inflicting dangerous stings. The bell of these small jellyfish is only ~20 mm (0.78 inch) wide with four long tentacles up to a metre long. The clinical presentation following a sting is known as the Irukandji syndrome, which can be severe, although generally not fatal when properly treated. The initial stinging sensation diminishes after a few minutes and you may not notice it. Subsequent localised development of limb pain is variable but occasionally quite severe. The sting area becomes reddened with small, gooseflesh-like lesions. An associated dry skin reaction may develop followed by excessive localised sweating. After 30 to 40 minutes following the sting, systemic symptoms usually commence, which may last 4 to 96 hours, but typically over 12. These are predominantly widespread pain (especially in the abdomen, large muscle groups, back and joints) often with severe headache. High blood pressure, rapid and irregular heartbeat, sweating, agitation, nausea and vomiting may also develop. As with box jellyfish, a sympathomimetic syndrome may develop. In severe cases, symptoms may progress to low blood pressure, pulmonary oedema, shock and

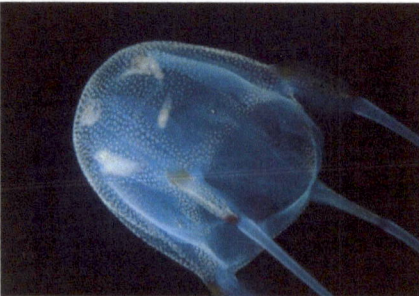

Distribution of Irukandji (*Carukia barnesi*).

heart failure. Heart attacks, even in the absence of recognised risk factors, may occur. Other stinging box jellyfish, in the same family as *Carukia* and *Malo,* have similar symptoms. They include several species in the genus *Carybdea* found throughout the southern Pacific Ocean.

The Indo-Pacific jellyfish, *Chiropsalmus quadrigatus,* is also commonly known as the box jellyfish but should not be confused with *Chironex fleckeri*. Therefore, to avoid confusion we use the common name false box jellyfish here. There is little specific information available for this species, but it is implicated in many serious stinging attacks and at least one death. A related species, *Chiropsalmus quadrumanus* (sea wasp), is a common jel-

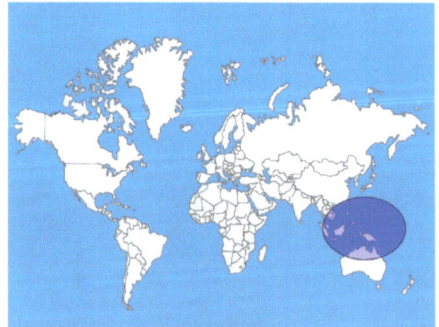

Box jellyfish (*Carybdea branchi*). Seascapeza/ Wikipedia. CC BY-SA 3.0. Available from https:// en.wikipedia.org/wiki/Carybdea_branchi#/media/ File:Carybdea_branchi9.jpg

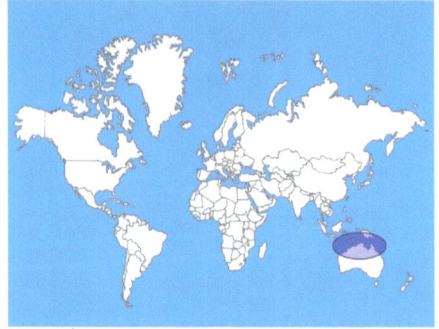

Distribution of Kingslayer (*Malo* spp.).

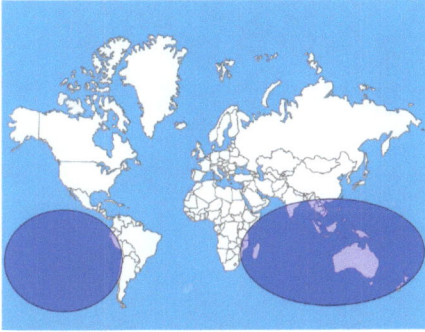

Distribution of box jellyfish (*Carybdea* spp.).

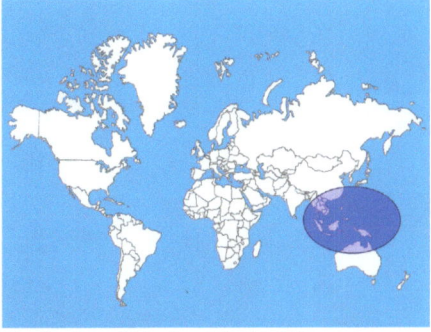

Distribution of False box jellyfish (*Chiropsalmus quadrigatus*).

lyfish in the western Atlantic Ocean and Gulf of Mexico and it has a potent sting.

The pink jellyfish, *Pelagia noctiluca*, is distributed in oceans worldwide appearing in abundance about every 10 to 12 years. During these cyclic proliferations, medusae swarms congregate near beaches with sig-

nificant envenomations. The stings usually produce minor cutaneous reactions such as reddened, inflamed and itchy eruptions, but some produce more dramatic lesions with some presenting as a burn-like response. A post inflammatory pigmentation may last several months, but this condition eventually resolves spontaneously.

The thimble jellyfish, *Linuche unguiculata*, ('sea lice') is a small, ~20 mm (0.78 inch) in diameter, species with a distinctive

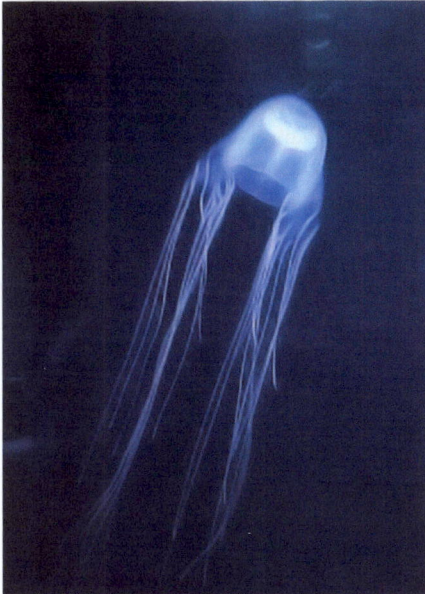

False box jellyfish (*Chiropsalmus quadrigatus*). Source: OpenCage/Wikipedia. CC BY-SA 2.5. Available from https://commons.wikimedia.org/wiki/File:Chiropsalmus_quadrigatus.jpg

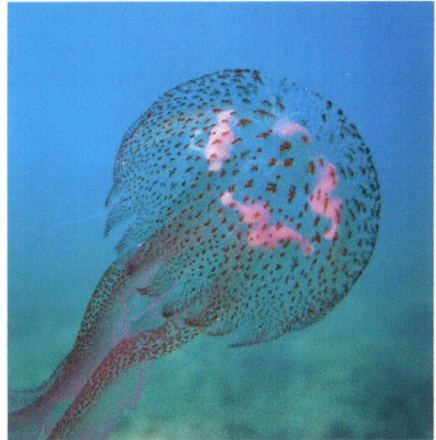

Pink Jellyfish (*Pelagia noctiluca*). Source: Alberto Romeo/Wikipedia, CC BY 3.0. Available from https://en.wikipedia.org/wiki/Pelagia_noctiluca#/media/File:Capo_Gallo_Pelagia_noctiluca.jpg

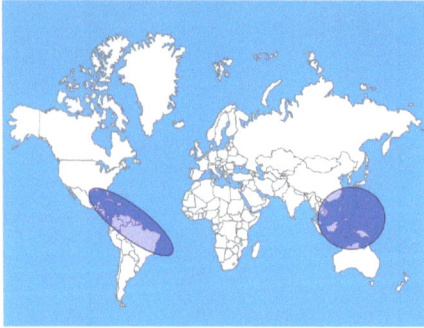

Distribution of Thimble jellyfish (*Linuche unguiculata*).

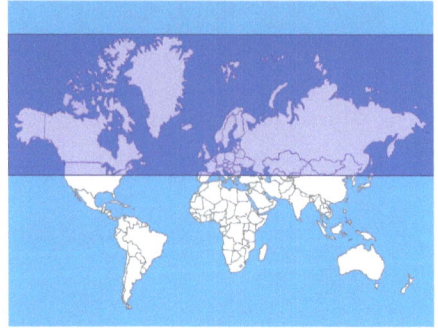

Distribution of Lion's mane jellyfish (*Cyanea capillata*).

dark brown internal appearance. They are widely distributed throughout the Caribbean and Gulf of Mexico and Indo-Pacific Ocean, breeding throughout the summer with populations usually peaking in May. They often form swarms of several thousand individuals, although isolated individuals are occasionally seen. Mild irritation follows an initial sting from this species. The site develops into an itchy, reddened rash within a few days known as seabather's eruption. Symptoms generally self-resolve, and antihistamines can reduce itching.

The lion's mane jellyfish, *Cyanea capillata*, is up to 2 m (6.5 feet) in diameter with long stinging tentacles and is distributed in cold boreal waters of the northern Atlantic and northern Pacific oceans and Arctic Sea. Their powerfully painful sting causes severe burning and blistering. Prolonged stinging events can cause muscle cramps, respiratory distress and, in some cases, death.

Portuguese man o'war (*Physalia physalis*) is a large jellyfish with up to 10 m (33 feet) long tentacles. It has a widespread distribution in mostly subtropical waters of the Atlantic, Pacific and Indian Oceans. Their large bluish gas filled sac (pneumatophore) moves them though the water similar to a sail, often towards shore. Man-o-war stings can be extremely painful and debilitating. Lesions and welts at the sting site are common. A red line develops with

Lion's mane jellyfish (*Cyanea capillata*). Source: Tim Gage/Flickr. CC BY-SA 2.0. Available from https://www.flickr.com/photos/timg_vancouver/9685088002/

Portuguese Man o'war (*Physalia physalis*). Source: NOAA

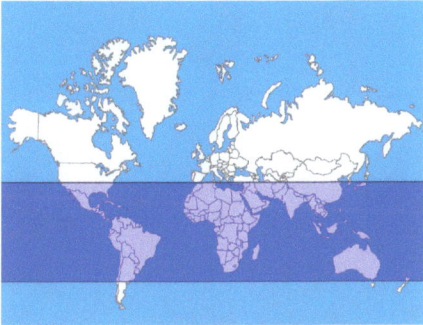

Distribution of Portuguese Man o'war (*Physalia physalis*).

Balloon corallimorph (Amplexideiscus fenestrafer). Source: Ahmed Abdul Rahman/Wikipedia. CC BY-SA 4.0. Available from https://en.wikipedia.org/wiki/Corallimorpharia#/media/File:Amplexidiscus_fenestrafer_Maldives.JPG

white lesions that may resemble a ladder-like pattern, and small wheals may develop resembling a string of beads. Most lesions dissipate within a few hours, but affected skin may remain reddened for about a day. Symptoms are highly variable, ranging from mild irritation and rashes to severe pain, systemic shock, and even death from respiratory failure. Persistent dull pain, often in the joints, follows the initial pain. A systemic syndrome lasting up to 24 hours may ensue with general discomfort, muscle cramps, headache, abdominal pain, chills, fever, nausea, vomiting, diarrhoea, hypotension (sometimes shock), nervousness, irritability, hysteria, confusion, abnormally fast heartbeat and cyanosis. This may require hospitalisation and possibly intensive care depending on systemic syndrome severity. Treatment is mainly supportive but may require advanced life support measures. Local corticosteroid creams may reduce inflammation.

Sea anemones
(Class Anthozoa)

An estimated 1000 sea anemone species occur in oceans globally. Fortunately, most sea anemones have weak nematocysts that

cannot penetrate human skin. However, some have dangerous, painful incapacitating stings. The balloon corallimorph (*Amplexidiscus fenestrafer*) is a colonial mushroom anemone distributed in the western Indo-Pacific Ocean. Its stinging tentacles can penetrate wetsuits causing significant pain. Initial symptoms vary from a prickly sensation to severe pain. The afflicted area can become red, swollen and blistered. The more dangerous anemones can cause shock and respiratory distress. Long-term neurological damage may occur. Treatment is the same as described earlier for jellyfish.

Unidentified sea anemone. Source: Bernard Spragg/Flickr. Public domain. CCO 1.0. Available from https://www.flickr.com/photos/volvob12b/14231557292/

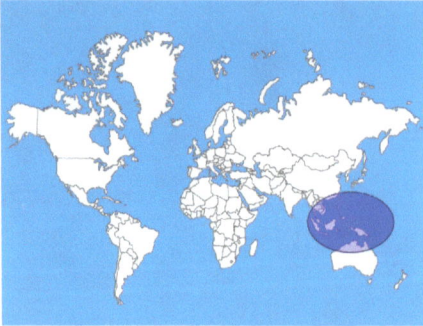

Distribution of Balloon corallimorph (*Amplexideiscus fenestrafer*).

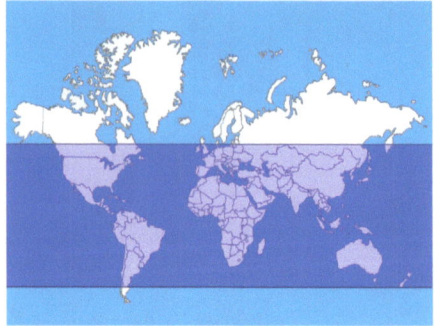

Distribution of Stinging sea ferns (*Aglaophenia* spp.).

Sea ferns
(Class Hydrozoa)

A few sessile, colonial hydroids can inflict dangerous stings, including the cypress sea fern, *Aglaophenia cupressina*, and white-stinging sea fern, *Aglaophenia philippina,* found in the central and southern Pacific Ocean. There are over 80 species assigned to this genus and it is possible that some of the other species may also be able to sting people. The slightest brush against their deceptively delicate fronds causes immediate pain. Each limb has rows of tiny nematocyst filled polyps. Stings begin as a patchy reddened area. Wheals may develop within 30 minutes and may take up to a month to heal. Effective pain relief can be obtained with a local anaesthetic ointment.

Corals
(Class Anthozoa)

Corals are a diverse group (over 2500 species) of sessile marine colonial coelenterates. About 1000 species secrete very sharp calcareous, rock-like shelters. These accumulate over time to create islands and atolls such as those on Australia's Great Barrier Reefs. They pose two potential threats: (1) both dead and living corals can produce serious cuts and scrapes, and (2) some living corals can inflict painful and potentially dangerous stings. About a dozen species of fire corals (*Millepora* spp.) are particularly noteworthy. They are commonly found in warm oceans

Stinging sea fern (*Aglaophenia cupressina*). Source: Bernard DuPont/Flickr. CC BY-SA 2.0. Available from https://www.flickr.com/photos/berniedup/8475698815/

Branching fire coral (*Millepora dichotoma*). Source: Tim Sheerman-Chase/Flickr. CC BY 2.0. Available from https://www.flickr.com/photos/tim_uk/10066860775/

worldwide. Stings burn and generally produce patchy, reddened, inflamed skin at the site of contact. Local anaesthetic ointment effectively relieves pain, but severe cases may require additional medical treatment. Corals are especially dangerous in the tropical Pacific Ocean. Coral danger signs posted on beaches should be heeded, and swimmers should take precautions to avoid accidental contact.

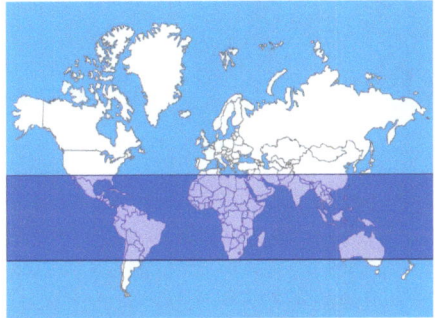

Distribution of Fire coral (*Millepora* spp.).

Bryozoans
(Phylum Bryozoa)

The ~4000 species of bryozoans are colonial animals found in both freshwater and marine environments. The sea chervil, *Alcyonidium diaphanum* (Family Alcyonidiidae), resembles seaweed and is widely distributed in the northern Atlantic Ocean. After repeated exposure, some (10%) of European coastal fishermen, especially in Denmark, England and France, develop an irritating, long lasting, allergic reaction commonly called Dogger's Bank itch (an erythematous dermatitis). The recommended treatment is normally a topical antihistamine. It is possible that some of the other dozen species of *Alcyonidium* distributed in oceans worldwide may produce similar contact dermatitis in people.

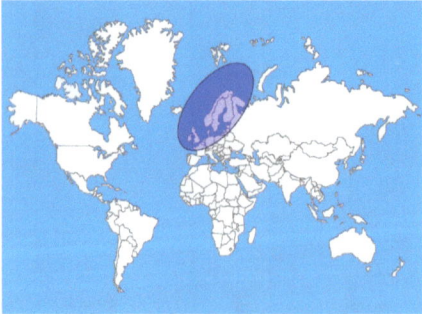

A marine bryozoan (*Alcyonidium diaphanum*). Source: Bernard Picton, National Museum Northern Ireland

A marine bryozoan (Alcyonidium diaphanum). Source: Bernard Picton, National Museum Northern Ireland

Distribution of marine bryozoan (*Alcyonidium diaphanum*).

Echinoderms
(Phylum Echinodermata: sea stars, brittle stars, sea urchins, sea cucumbers)

The 6000 species of echinoderms are exclusively marine, mostly bottom dwellers, and perhaps the iconic symbol of sea life. Overlapping calcareous plates called ossicles form a tough skeleton with projecting spines, thus the name echinoderm or 'spiny skin.' This skeleton often forms articulating rays or flexible arms such as those of sea stars and brittle stars, or rigid structures in sea urchins and sand dollars. Certain sea stars, brittle stars and sea urchins are toxic and/or cause physical damage.

Crown of thorns starfish (*Acanthaster planci*) with darker colour form. Source: US National Park Service

Sea stars and brittle stars
(Class Stelleroidea)

There are some 1500 species of sea stars (starfish) that inhabit coastal waters worldwide. They are often colourful and can be

Crown of thorns starfish (*Acanthaster planci*) with lighter colour form. Source: US National Park Service

seen crawling about rocks and on muddy or sandy bottoms. Most are harmless, but the distinctive tropical Indo-Pacific crown of thorns (*Acanthaster planci*) starfish is venomous. It has more than a dozen spiny arms and grows to more than 30 cm (1 foot) in diameter. It's sharp, venomous spines can penetrate gloves, boots and wetsuits causing severe pain, swelling, profuse and frequent vomiting, numbness and occasionally paralysis. Immersion of an afflicted body part in hot water, when practical, can help reduce pain, but pain may last several days. Severe envenomation requires medical attention.

The mosaic sea star (*Plectaster decanus*), distributed off the southern coast of Australia, can cause a skin rash if handled. The chain-link brittle star (*Ophiomastix*

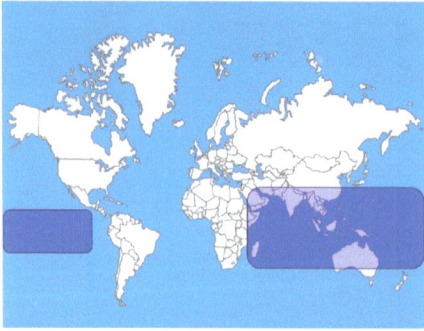

Distribution of Crown of thorns starfish (*Acanthaster planci*).

Chain-link brittle star (*Ophiomastix annufosa*). Source: Frédéric Ducarme/Wikipedia. CC BY-SA 4.0. Available from https://sv.wikipedia.org/wiki/Ophiomastix_annulosa#/media/File:Ophiomastix_annulosa.JPG

annufosa, Class Ophiuroidea) has caused deaths of small animals. Contact with these species should be avoided.

Mosaic sea star (*Plectaster decanus*). Source: Richard Ling/Wikipedia. CC BY-SA 2.0. Available from https://en.wikipedia.org/wiki/Plectaster_decanus#/media/File:Plectaster_decanus.jpg

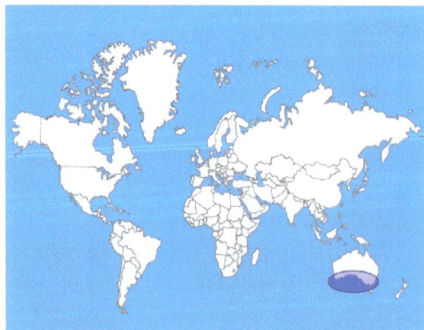

Distribution of Chain-link brittle star (*Ophiomastix annufosa*).

Sea urchins
(Class Echinoidea)

Similar to starfish, some sea urchins are venomous. Some 950 species of sea urchins inhabit oceans worldwide. They have venomous spines and/or minute stalked appendages (pedicellariae). The spines are brittle and can break-off in your skin causing additional physical trauma. Urchins in the Family Toxopneusidae are especially noteworthy for their venomous properties. They have short thick spines projecting outward through a flower-like display of pedicellariae, which

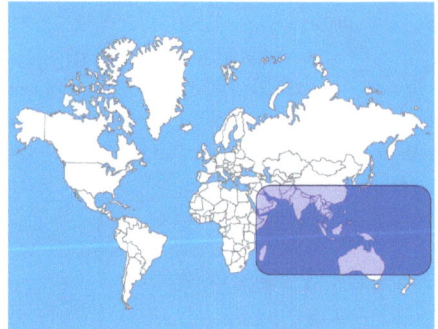

Distribution of Mosaic sea star (*Plectaster decanus*).

Sea urchin. Source: NOAA

bear hook-like venomous jaws. Envenomation may cause general discomfort, nausea, vomiting, diarrhoea, headaches, intense pain, respiratory distress, paralysis and occasionally death. Hot water immersion (tolerable to touch) may help reduce symptoms. Even with hot water treatment, the victim may need professional medical attention because the spines continue releasing venom until they are removed, which may require excision in severe cases. Then, in addition, follow up local antibiotic therapy may be needed to reduce secondary infection risk as well as local anaesthesia to alleviate pain.

Sea cucumbers
(Class Holothuroidea)

Sea cucumbers are our final, although rarely, dangerous echinoderm group. There are 900 species distributed in oceans worldwide. Some defensively eviscerate their intestines and extrude white sticky threads when threatened or handled. These threads and excreted mucus contain the toxin holothurin, which can cause skin and eye irritation including burning, inflammation, redness, intense pain, eye damage and sometimes blindness. Some also ingest other stinging animals incorporating and excreting those toxins defensively. Most toxicity problems are due to consuming the flesh of poisonous species, which can be deadly in some situations.

Three-rowed sea cucumber (*Isostichopus badionotus*). Source: Roban Kramer/Flickr. CC BY-SA 2.0. Available from https://www.flickr.com/photos/robanhk/3370265047/

Sea cucumber (left) and sea urchin (right). Source: Ratha Grimes/Flick. CC by 2.0. Available from https://www.flickr.com/photos/ratha/8491558858/

Unknown species of sea cucumber. Source: Thomas Quine/Flickr. CC By-SA 2.0. Available from https://www.flickr.com/photos/quinet/5402056393/

Segmented worms (Phylum Annelida: leeches, polychaetes)

Annelid worms are the segmented worms. Most, such as the familiar earthworm, are harmless. However, there are a few that can hurt you or cause discomfort and annoyance.

Leeches
(Class Hirudinea)

There are ~500 known species of leeches. The majority inhabit diverse standing freshwater habitats worldwide, but some are marine. Others occupy moist, temperate and tropical terrestrial environments. Leeches have 34 body segments, although their form and structure varies widely. Land leeches can be large in size. The largest, Amazon leech (*Haementeria ghilianii*), can grow to over 45 cm (18 inches) long. Most species are free-living scavengers or predators, but some suck the blood from vertebrate hosts, including humans. Leeches often attach to the skin where they can be removed relatively easily by gently placing a sharp edged object, such as a fingernail, under the feeding sucker until it releases. In some circumstances, they have been known to invade the sinuses, which may require surgical extraction. The European medicinal leech (*Hirudo medicinalis*), Asian medicinal leech (*Hirudinaria manillensis*) and the giant Amazon leech (*Haementeria ghilianii*) are common examples. The land leeches (*Haemadipsa* spp.) of East Asia and South-East Asia are

European medicinal leech (*Hirudo medicinalis*). Source: Pixabay. CC0. Available from https://pixabay.com/en/leech-leech-therapy-medicinal-leech-1055446/

all blood feeders. Should your plans include visiting East and South-East Asian jungles, you will find blood-sucking terrestrial and aquatic leeches very annoying.

Leeches respond rapidly to the presence of a potential host whether it is in water or

Asian land leech (*Haemadipsa picta*). Source: Gunther Eichhorn/Wikipedia. CC BY-SA 3.0. Available from https://en.wikipedia.org/wiki/Haemadipsa#/media/File:Haemadipsa_picta.jpg

Land leech (*Haemadipsa zeylanica japonica*). Source Pieria/Wikipedia. Public domain. Available from https://upload.wikimedia.org/wikipedia/commons/f/f9/Haemadipsa_zeylanica_japonica.jpg

Distribution of Amazon leech (*Haementeria ghilianii*).

on land. They are stimulated to seek a host based on physical disturbance of the habitat, but body odour, movement, carbon dioxide, and perhaps other factors may similarly arouse them to stimulate host seeking and attachment behaviours. Once a host is located, they immediately begin searching for an attachment site, and attachment may take only 20 to 30 seconds. They attach to the host with an anterior sucker through which feeding also occurs. If a favourable attachment site is not found on the first try, leeches may release and move to another site to repeat exploring until finding a suitable attachment point.

Once land leeches are attracted they may display a characteristic 'stand up' behaviour where they reach and sway, thus appearing to 'wave.' Upon the slightest contact, they then adhere and begin 'exploring' for an attachment site. Leeches may enter any variety of openings on clothing or shoes such as a tear, loose zipper, boot eyelets, or even loosely woven fabric. They also are known to crawl in shoes and penetrate your sock to feed. You may not notice the bite until you remove your clothing or boots, or you get a 'wet' feeling due to the spilled blood.

Both aquatic and land leeches quickly perforate your skin using three cutting plates each bearing minute teeth that

Distribution of European medicinal leech (*Hirudo medicinalis*).

Distribution of Asian land leech (*Haemadipsa picta*).

protrude into the anterior sucker cavity. In order to keep the host's blood from clotting as they feed, leeches secrete salivary fluid containing hirudin or hemetin anticoagulants (depending on the species). Anticoagulants allow them to feed freely until engorged. Blood may continue to flow from the wound up to an hour after they drop or are removed. A bite wound may ooze for up to 5 hours, thus giving the unknowing victim a wet feeling. Leeches can ingest a surprising amount of blood in minutes, up to several times their own weight, which can make for relatively short attachment times. Because they often ingest so much blood, and have low rates of digestion, they may go months between blood meals. Although excessive blood loss due to multiple feeding events may weaken the victim, the most significant threat posed to people by these worms is arguably psychological distress. However, delayed irritation and itching at bite wound, and secondary infections also may occur. When leeches are removed, the wound should be thoroughly washed and use alcohol or similar disinfectant solution to prevent infection and help reduce further blood loss.

Polychaetes or bristle worms
(Class Polychaeta)

This large group of annelid worms contains over 5300 species. Most are less than 10 cm (4 inches) long and 2 to 10 mm (0.08 to 0.40 inch) wide, although one species of Eunice may grow to 3 m (10 feet) long. Certain free-living marine fire worms (*Hermodice carrunculata*, *Eurythoë* spp.) have retractable, hollow, toxin-filled setae they can extend when threatened. These setae break when they enter the skin, thereby inflicting a painful sting. Reactions may include a

Bearded fire worm (*Hermodice carunculata*). Source: Prilfish/Flickr. CC BY 2.0. Available from https://www.flickr.com/photos/silkebaron/21130025292/

burning sensation lasting hours to days, local inflammation, itching, numbness, and possibly secondary infection. To avoid aggravating the stings, use forceps or sticky tape to remove the embedded setae before you further treat the wound. Use isopropyl alcohol, vinegar or diluted (10%) ammonia to soak the affected area to relieve pain. A topical anaesthetic such as benzocaine also can be used to relieve pain. A topical antibiotic ointment may help prevent a secondary infection.

Other marine polychaetes use a fang tipped eversible proboscis (rather than stinging setae) to inflict painful, venomous bites. This group includes *Eunice aphrodi-*

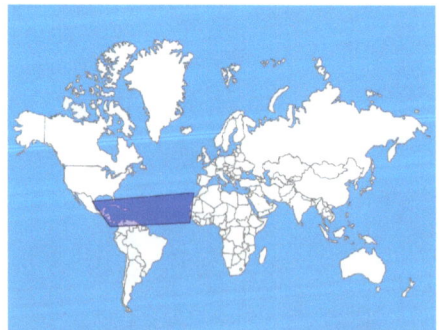

Distribution of bearded fire worm (*Hermodice carunculata*).

Bristle worm (*Eurythoe complanata*). Source: Brocken Inaglory/Wikipedia. GNU FDL. Available from https://commons.wikimedia.org/wiki/File:Eurythoe_complanata.jpg

Marine polychaete (*Glycera alba*). Source: Hans Hillewaert/Wikimedia. CC BY-SA 4.0. Available from https://commons.wikimedia.org/wiki/File:Glycera_alba_(dim).jpg

tois, *Glycera* (~75 species) *Onuphis* (~80 species). All are widely distributed in oceans worldwide. The bites of *Glycera* (~75 species) are reportedly similar to wasp stings. Typically, the victim experiences pain, swelling, redness and itching that may last several days. Small lesions at the bite site may appear, but recovery is otherwise uneventful.

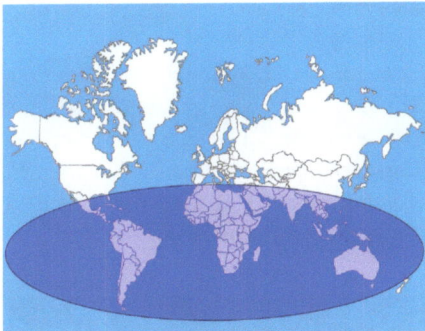

Distribution of bristle worm (*Eurythoe complanata*).

A marine polychaete (bobbit worm, *Eunice aphroditois*) in its burrow. Source: Rickard Zerpe/Flickr. CC BY-SA 2.0. Available from https://www.flickr.com/photos/krokodiver/14217679648/

Molluscs
(Phylum Mollusca: cephalopods, cone shells, sea butterflies, sea slugs, snails)

The molluscs include over 70 000 described terrestrial, freshwater and marine species that are distributed worldwide. However, only a few marine and freshwater species pose a threat to human health through injury, illness or envenomation.

Cephalopods
(Class Cephalapoda)

These creatures are exclusively marine organisms that inhabit oceans worldwide and are commonly known as squids, octopuses and cuttlefishes. There are ~600 known species. Most are small to moderate sized (up to 6–70 cm or 28 inches), but the largest living invertebrates are cephalopods. For example, the giant squid can grow to over 16 m (52 feet). Cephalopods have a hardened tooth-like beak and toxic salivary juices that are used to kill or immobilise prey. Therefore, some cephalopods can pose a threat to people who encounter them or attempt to handle them because they can use their beak to inflict painful and potentially dangerous bites. Additionally, there may be nausea, vomiting, diarrhoea, fever, headache, chills and respiratory distress. Some bite victims have experienced blurred vision, difficulty talking, convulsions, numbness of the extremities and paralysis. In most cases, complete recovery normally occurs within

Cuttlefish. Source: Sylke Rohrlach/Flickr. CC BY-SA 2.0. Available from https://www.flickr.com/photos/87895263@N06/8428878034/

2 days. However, although rare (less than 1%), bites from some species can be fatal. For example, blue-ringed octopuses (*Hapalochlaena maculosa, H. lunulata, H. fasciata*), are small (up to ~20 cm, 8 inches), non-aggressive animals, but are capable of

Greater blue-ringed octopus (*Hapalochlaena lunulata*. Source: Angell Williams/Flickr. CC BY 2.0. Available from https://www.flickr.com/photos/53357045@N02/4973031885/

Southern blue-ringed octopus (*Hapalochlaena maculosa*). Source: Bernard DuPont/Flickr. CC BY-SA 2.0. Available from https://www.flickr.com/photos/berniedup/8466102934/

inflicting a lethal bite. A single blue-ringed octopus may possess enough tetrodotoxin, a potent neurotoxin, to kill 10 people. However, only about a dozen human deaths have been attributed to this species.

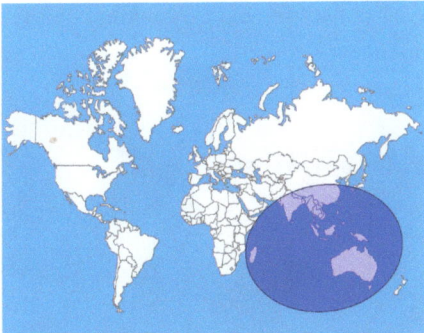

Distribution of blue-ringed octopus (*Hapalochlaena* spp.).

Giant Pacific octopus (*Enteroctopus dofleini*). Photo: David E. Bowles

Gastropods
(Class Gastropoda)

There are 35 000 described gastropod species including marine cone snails, pteropods, nudibranchs and freshwater snails. This group contains some species that have substantial potential to harm people.

Cone shells
(*Genus* Conus)

This is the most dangerous group of gastropod that has been implicated in human injuries. There are over 500 species distributed in temperate to tropical regions of oceans worldwide. Some have inflicted serious and sometimes lethal stings. The most dangerous species are distributed in the western and south-western Pacific from Japan to Australia, including the Loyalty Islands, the New Hebrides, New Britain, the Seychelles, New Caledonia, New Guinea and the Paumotou Islands.

Cone shells are predators. They have a long, highly manoeuverable proboscis. The proboscis has a detachable venom gland equipped with a sharp tooth. Cone shells lie buried in the sand and use this tooth proboscis apparatus to harpoon fish and other creatures, which they can do quite rapidly. This dart-like tooth operates faster than people can react, so you should never pick-up or handle a cone shell without hand protection. Stepping on a cone shell also can result in a sting. The victim typically only receives a small puncture wound, usually on the hand, that results in an immediate sharp, burning pain. Alternatively, the victim may experience a rapidly spreading numbness, rather than intense pain, that radiates from the puncture wound. Additional symptoms may include a rapidly developing flaccid paralysis of the

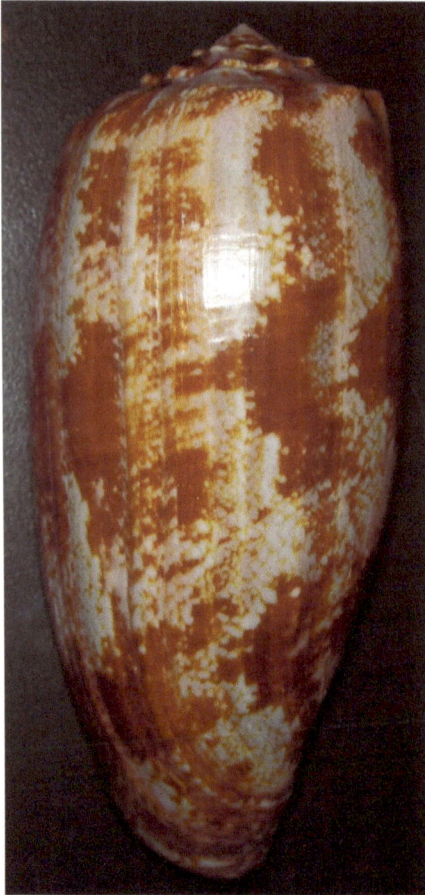

Textile cone (*Conus textile*) with proboscis extended. Source: Harry Rose/Flickr. CC BY 2.0. Available from https://www.flickr.com/photos/macleaygrassman/9271210509/

That said, when fatalities do occur, death typically takes less than 12 hours, but can occur in 6 hours or less. Therefore, any sting from a cone shell should be considered potentially life threatening, and the

Geography cone (*Conus geographus*). Source: James St. John/Flickr. CC BY 2.0. Available from https://www.flickr.com/photos/jsjgeology/24422159755/

afflicted limb, impaired speech, blurred vision, loss of sensation and complete absence of reflexes. Most victims do not experience any respiratory difficulty, but there may be swelling and skin discolouration, particularly in the area of the sting. In most cases, symptoms resolve within 24 hours, but complete recovery can take hours to weeks, depending on the species, venom and amount of venom delivered.

Cone shell (*Conus* sp.). Photo: David E. Bowles & Mark Pomerinke, United States Air Force

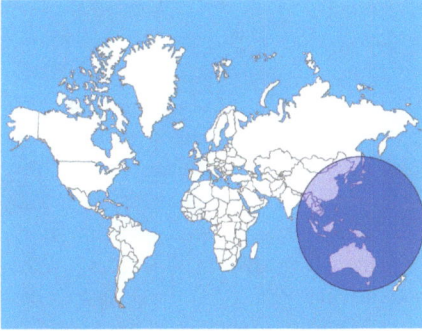

Distribution of Cone shell (*Conus* spp.). Only distribution of the most dangerous species is shown.

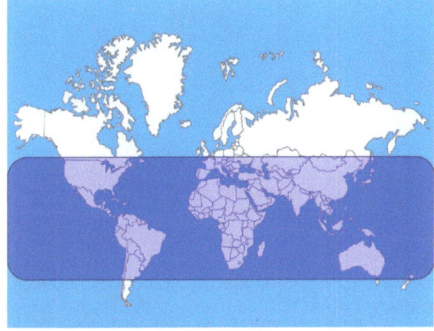

Distribution of sea butterfly (*Creseis clava*).

victim should seek immediate medical attention. The most commonly implicated cone shell in fatal stings is the geography cone, *Conus geographus*, although others also are capable of inflicting serious injury.

Sea butterflies (pteropods) (Order Thecosomata)

The 100 or so sea butterfly species are small snails (<5 mm) inhabiting oceans worldwide. Unlike most snails, they are capable of sustained swimming because their body is modified into two flapping wings, thus the common name. Sea butter-

A sea butterfly. Source: United States Geological Survey/Wikipedia. Public domain. Available from https://upload.wikimedia.org/wikipedia/commons/a/aa/Sea_butterfly.jpg

flies are primarily a nuisance rather than a serious health threat. Contact with them while participating in water sports and other activities may result in irritating stings, which subsequently may cause a raised (maculopapular) rash. The rash generally self-resolves within a few days with no other serious symptoms. The sea butterfly, *Creseis clava,* is distributed worldwide in warm seas and it has been implicated as a nuisance species because it can sometimes occur in high densities.

Sea slugs (Nudibranchs) (Order Nudibranchia)

Many of the over 2300 recognised sea slug species worldwide are spectacularly beautiful. They are often brightly coloured (red, yellow, orange, blue, green and combinations), which is nature's warning 'do not touch'. Numerous species in the genera *Aeolidia*, *Glaucus* and *Hermissenda* feed on coelenterates (see earlier) and absorb their stinging nematocysts for their own defensive purposes. The blue sea slug, *Glaucus atlanticus*, inhabits open water where it generally floats on the surface and feeds on the Portuguese man o'war. Nudibranch stings are essentially like those of the

Nudibranch (*Hermissenda crassicornis*). Source: Ed Bierman/Flickr. CC BY 2.0. Available from https://www.flickr.com/photos/edbierman/4629351946/

Shag-rug nudibranch (*Aeolidia papillosa*). Source: Jeffy Kirkhart/Flickr. CC by 2.0. Available from https://www.flickr.com/photos/jkirkhart35/5843448733/

Blue sea slug (*Glaucus atlanticus*). Source: Taro Taylor/Wikipedia. CC BY 2.0. Available from https://en.wikipedia.org/wiki/Glaucus_atlanticus#/media/File:Glaucus_atlanticus_1_cropped.jpg

coelenterates themselves. Refer to the coelenterate section for similar symptoms, outcomes and treatment. Nudibranchs also possess a wide range of chemical defences to deter would be predators. Among these, some species are capable of secreting sulfuric acid via the mucous from skin glands, which can burn the skin upon contact. Their diverse coloration makes specific identification difficult so it is best to avoid all contact with nudibranchs.

Freshwater snails (Orders Mesogastropoda and Basommatophora)

Several freshwater snail genera are intermediate hosts for trematode flatworms (flukes). These snails like slow moving or stagnant water where they feed on organic waste and aquatic vegetation. The various flukes have complex life cycles but the juvenile flukes, or cercaria, which normally parasitises other animals such as wading birds, can infect people. The cercaria leaves the snail during peak daylight hours and seeks out the host for their next life stage. Most often, the definitive hosts are birds, but when humans enter water that is infested with such parasites, they can become accidental hosts. The cercariae contact the host and burrow into the skin, which later produces pustules that itch intensely. Such infestations do not have any long-term consequences nor do they pose a disease threat. Such benign infestations are called 'swimmer's itch (schistosome cercarial dermatitis).' Swimmer's itch also can be contracted in some brackish and marine waters. In other areas, snails harbour flukes that cause a very serious disease, schistosomiasis (bilharziasis), which affects over 240 million people worldwide.

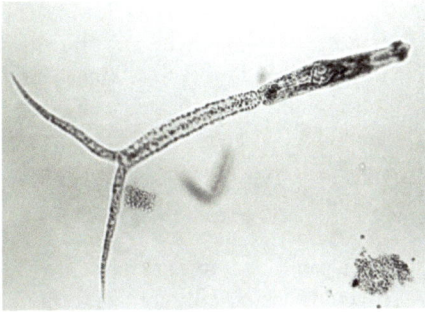

Typical fluke cercaria that causes swimmer's itch. Source: CDC, Minnesota Department of Health/ Wikipedia. Public domain. Available from https:// en.wikipedia.org/wiki/Swimmer%27s_itch#/media/ File:Schistosomal_cercaria.jpg

Risk area for schistosomiasis worldwide.

Like the causative flukes, human schistosomiasis is a complex disease involving several snail genera and eight species of flukes (*Schistosoma*), which are widely distributed. Initial symptoms of schistosomiasis include skin rash, fever, headache, muscle ache, bloody diarrhoea, cough, malaise and abdominal pain, which usually appear within days to weeks after exposure. Without treatment, chronic schistosomiasis develops once the flukes release eggs that damage abdominal organs, female genital organs, heart, lungs and sometimes the brain. Irreversible damage, including liver cancer, may result. When in the known range of schistosomiasis, an experience with 'swimmers itch' or other initial symptoms after wading, swimming, washing, drinking or bathing, or eating unsanitised raw foods, should be cause for concern and medical attention should be sought. A commonly available treatment for schistosomiasis is Praziquantel, an antihelminthic drug. Because the snails that harbor the schistosomiasis parasite are difficult to identify and are widely distributed, the risk is better presented in terms of the distribution of the disease itself.

The freshwater snail genera harbouring schistosomiasis cercaria include *Bulinus*, *Biomphalaria* and *Oncomelania*. Snails in the genus *Bulinus* are distributed throughout Africa, the Middle East, Madagascar, Mauritius and India. *Australorbis* spp. (formerly *Biomphalaria* spp.) occurs in Africa, the Middle East, Madagascar, South America and some Caribbean Islands. Some species also occupy the southern United States southward through Central America, but those species of *Australorbis* do not carry schistosomes. *Oncomelania* is widely distributed throughout Asia including China, Japan, the Philippines and Sulawesi (Celebes). *Oncomelania* snails are amphibious, readily climb streamside vegetation, and can be found on moist soils near water. The WHO International Association for Medical Assistance to Travellers (IAMAT) site (https://www.iamat.org) provides an excellent snail identification guide, but accurate identification is difficult for the layperson. Although there are several schistosomiasis eradication programs that have had varying degrees of success, we still recommend using appropriate caution

A snail (*Biomphalaria glabrata*) that can transmit schistosomiasis. Source: Fred A. Lewis, Yung-san Liang, Nithya Raghavan & Matty Knight/Wikipedia. CC BY 2.5. Available from https://en.wikipedia.org/wiki/Biomphalaria#/media/File:Biomphalaria_glabrata.jpg

throughout its historic range. We recommend travellers use the following guidelines to minimise the risk of contracting schistosomiasis in areas where it is endemic:

Snail (*Bulinus truncatus*). Source: Fred A. Lewis, Yung-san Liang, Nithya Raghavan & Matty Knight/Wikipedia. CC BY 2.5. https://commons.wikimedia.org/wiki/File:Bulinus_truncatus.jpg

1. Avoid contacting slow flowing or standing fresh water. This includes wading, swimming, drinking, hand washing and other such contacts.
2. Wear high, waterproof boots or hip waders if you must enter streams or swamps.
3. Use rubber or latex gloves if you must place your hands in such water.
4. If you do accidentally contact freshwater, immediately and vigorously scrub that area of skin with rubbing alcohol, and dry with a towel to reduce the possibility of infection.
5. Never drink water from a natural source such as a stream because cercariae can burrow through mouth mucosa. Only drink bottled, purified water and make sure all bottles are sealed.
6. If you must drink water from a natural source as an emergency ration, boil it for a minimum of 5 minutes, or treat it with chlorine tablets following label directions. An approved water filtration/purifier device may also be safely used.
7. Avoid salads and raw vegetables that may have been washed in contaminated water.

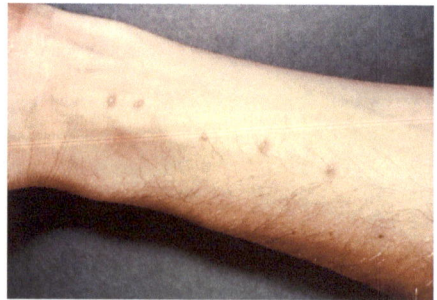

Blisters caused by Schistosomes entering skin. Source: US Centers for Disease Control and Prevention/Wikipedia. Public domain

Spiders
(Order Araneae)

About 25 000 species of spiders are known worldwide. Although most spiders are completely harmless or cause only mild irritation with their bites, the bites of a few can cause substantial pain, suffering, and even death. Spiders as a group, including the potentially dangerous species, are shy and secretive and most human encounters with them are accidental. Although only a few species are considered dangerous, accurate identification of the species can be difficult. Therefore, it is best to avoid contact with spiders unless you are certain of their identity and relative threat level. Appendix 6 summarises the most dangerous spiders and their distribution. We highlight a few of the most dangerous spiders and then cover prevention and treatment, including first aid measures.

Wandering/banana spiders
(Genus *Phoneutria*)

Some South America banana spiders (*Phoneutria fera, Phoneutria ochracea, Phoneurtria* spp.) are aggressive and known to envenomate humans, even causing death. These South American spiders are commonly called wandering spiders, but they are not closely related to the African wandering spiders. Hundreds of people are bitten by wandering spiders yearly, mostly during winter months. The bites are painful, becoming deeply seated and generalised with swelling around the bite site after a few hours. The venom is a potent neurotoxin affecting both the central and

Wandering spider (*Phoneutria fera*). Source: Bernard DuPont/Flickr. CC BY-SA 2.0. Available from https://www.flickr.com/photos/berniedup/10623228224/

peripheral nervous system. Symptoms are variable, including altered pulse rates, irregular heartbeat, temporary blindness, sweating, fever and increased glandular functions, especially the kidneys. Roughly, 24 hours following the bite, the victim may suffer general muscle pains and prostration. Fatalities are not common, and children aged under 6 are most vulnerable. Bite victims should seek prompt medical treatment. No antivenom is available.

Distribution of Wandering spiders (*Phoneutria* spp.).

Widow spiders
(Genus *Latrodectus*)

Widow spiders are among the most dangerous spiders to humans. Although timid and reclusive, they can inflict painful and potentially deadly bites when they are provoked or if you contact them accidentally. Approximately 40 black widow species occur worldwide, but because the species can be difficult to distinguish we here consider all members of this genus to be dangerous. Known medically important species occur in the Middle East, Europe, Madagascar, Africa, Asia, Australia and throughout the Western Hemisphere. Colouration and markings among the species of widow spiders varies broadly. Most are ~25 mm (1 inch) long and have large, shiny, hairless, bulbous abdomens and blackish colouration. Most have variously shaped red markings on their abdominal venter (bottom), and some may have similar dorsal (top side) markings. The most well known markings are the southern black widow's (*Latrodectus mactans*) red hourglass in North America and the Austro-Asian region redback's (*Latrodectus hasselti*) red dorsal spot or stripe. Geographically unique common names for *Latrodectus* species include: the Mediterranean black widow, malmignatte or karakurt (*Latrodectus tredecimguttatus* in Eurasia); shoe-button spider (South Africa); katipo (New Zealand); and redback (Australia). Other medically important species of widow spiders are not all black and include the brown widow (southern US, tropical areas worldwide), red widow (central and southern Florida, Africa), and northern black widow (northern Florida to Canada).

Because of their secretive habits, widow spider bites are relatively rare, and the toxicity of their neurotoxic venom varies widely among species. The severity of the bite and envenomation also varies widely among the species. The bites of some (e.g. black widow, red-back spider) can be severe, while those of others (e.g. brown widow) are milder. Initially *Latrodectus* bites may be painless and only involve mild dermatologic responses including localised redness at the bite wound, sweating, and erect or bristled hair within the first half hour. The lymph nodes draining the wound site may become swollen, palpable and painful. In addition, cyanosis may develop around the bite site and skin hives or itchy wheals may erupt. Sometimes significant systemic symptoms may ensue within minutes following a bite, resulting in the clinical condition latrodectism. Latrodectism's most significant feature is severe localised pain that increasingly intensifies to generalised, persistent pain accompanied by severe muscle pain, rigid (board-like) abdominal cramping, tightness through the chest, difficulty breathing and nausea. Although rare, 4 to 5% of envenomated victims die without treatment following such symptoms. Finally, widow spider bites are occasionally misdiagnosed as ruptured ulcers, acute appendicitis, kidney problems or food poisoning.

To help someone bitten by a widow spider, you should immediately perform first aid and seek medical treatment. Hospitalisation may be necessary in some cases. High-risk patients may require antivenom, which is commercially available (Merck®) for the North American black widows (*L. mactans, L. indistinctus*), the Australian red-back spider (*L. hasselti*), the South African brown widow (*L. geometricus*), the Argentinian *L. mactans*, and the Mexican widow spider. European widow (*L. tredecimguttatus*) antivenom is no

Southern black widow (*Latrodectus mactans*).
Source: David E. Bowles

longer produced. Laboratory studies show these antivenoms are effective and have cross-species reactivity. They produce few allergic responses, although some individuals may be sensitive to them.

Web of southern black widow (Latrodectus mactans) showing irregular pattern of silk threads. Source: David E. Bowles & Mark Pomerinke, United States Air Force

Brown widow (*Latrodectus geometricus*). Source: Matthew Field/Wikipedia, GNU FDL. Available from https://upload.wikimedia.org/wikipedia/commons/d/d7/Brown_widow_spider_Latrodectus_geometricus_low_oblique_view.jpg

Red widow (*Latrodectus bishopi*). Source: Florida Division of Plant Industry Archive/Wikipedia. CC BY 3.0. Available from https://en.wikipedia.org/wiki/Latrodectus_bishopi#/media/File:Latrodectus_bishopi.jpg

Redback spider (*Latrodectus hasseltii*). Source: Toby Hudson/Wikipedia. CC BY-SA 3.0. Available from https://en.wikipedia.org/wiki/Redback_spider#/media/File:Latrodectus_hasseltii_close.jpg

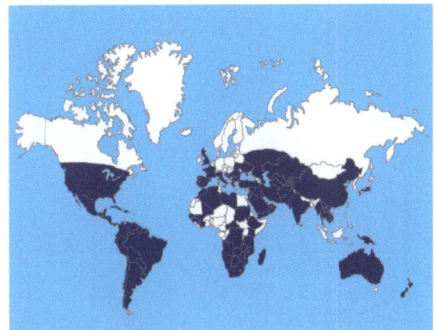

Distribution of Widow spiders (*Latrodectus* spp.).

Brown recluse or fiddleback spiders
(Genus *Loxosceles*)

Over 100 described species of spiders in the genus *Loxosceles* are distributed worldwide. Most of these species probably are not dangerous or their status is unknown, but some pose a potential health threat. For this reason, we consider all members of this genus to be potentially dangerous. All species of *Loxosceles* have a fiddle-shaped mark on the cephalothorax, legs long and sleek, and brown to grey colouration. However, the fiddle-shaped mark is not well defined in some species. *Loxosceles reclusa*, a North American species, is perhaps the most recognised *Loxosceles* species. Because these shy spiders tend to hide among household goods, they have on occasion been transported around the world where some were introduced at the new locations. For example, the Chilean recluse, *Loxosceles laeta*, and the Mediterranean recluse, *Loxosceles rufescens* were

Brown recluse spider showing fiddle mark on cephalothorax. Source: Mike Keeling/Flickr. CC BY-ND 2.0. Available from https://www.flickr.com/photos/pachytime/3166642926/

introduced to the United States and Canada, and likely elsewhere. Some of these introduced populations were exterminated before they expanded, but others are almost certainly naturalised and continue to exist.

Considerable myth and misinformation surround brown recluse spiders and the relative seriousness of their bites. Documented bites are rare relative to their population densities in human inhabited structures. Furthermore, many 'brown recluse' bites reported by medical practitioners are misdiagnosed, and are really due to other factors. Generally, recluse spiders are shy and reclusive, but they will

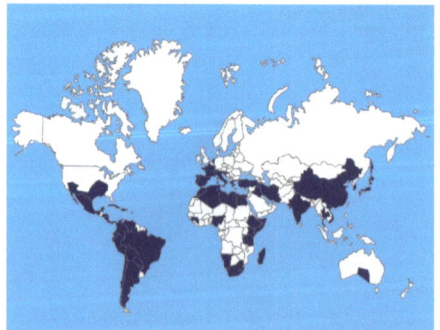

Brown recluse spider (*Loxosceles reclusa*). Source: Alex Wild/Flickr. Public domain. Available from https://www.flickr.com/photos/131104726@N02/25823605780/

Distribution of fiddleback spiders (*Loxosceles* spp.).

bite if harassed or accidentally contacted. In such cases, multiple bites may occur. Most bites are asymptomatic or self-resolving, but reactions can vary from a small pimple-like lesion, to severe, tissue necrosis and ulcers. The necrotising venom is potent and has the potential to destroy living tissue.

Bites initially produce little pain. In certain cases, the reaction following a bite progresses to a light-coloured halo around the bite with a greyish-blue centre surrounded by red. A typical lesion is around 15 mm (1/2 inch) or smaller with raised edges and a sunken centre, but ~12 to 24 hours following envenomation, the bite site becomes painful, swollen and red or mottled purplish-red. Later, hardened tissue, a blister or pimple may also form, followed by necrosis. This response is termed necrotising arachnidism syndrome and it occurs in only ~10% of bite victims. Severe necrotic wounds may take months to heal or require surgical intervention. Rarely, additional systemic reactions occur including formation of blood clots throughout the body's small blood vessels (disseminated intravascular coagulation), passing bloody urine, acute kidney failure,

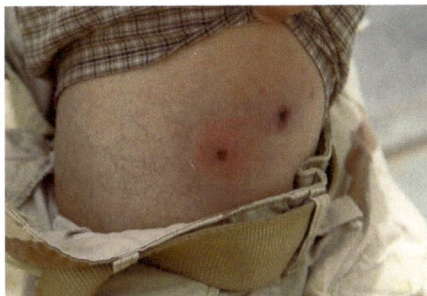

Mild necrotic lesion due to suspected brown recluse bite. Source: James A. Swaby

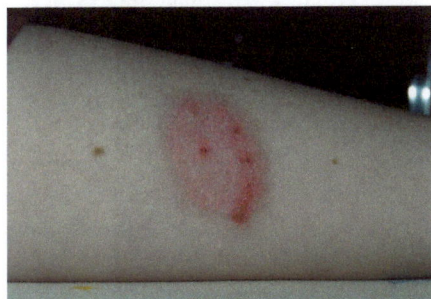

Necrosis following a suspected brown recluse bite. Source: S. J. Pyrotechnic/Flickr. CC BY-SA 2.0. Available from https://www.flickr.com/photos/sarahakabmg/111172858/

convulsions, coma and rarely death. Antivenom for *Loxosceles* is available in some countries, but there is little evidence to support its effectiveness, particularly against local effects. If a bite from a recluse spider is confirmed, the victim should immediately seek first aid and medical attention.

Funnel web spiders
(Genera *Atrax* and *Hadronyche*)

There are roughly 35 known species of Australian funnel web spiders. The genus *Atrax* is known only from New South Wales, Australia, while *Hadronyche* is found throughout eastern Australia and Tasmania. *Atrax robustus* and *Hadronyche formidabilis* are arguably the most venomous and dangerous spiders in the world. *Atrax robustus* (Sydney funnel web spider) is restricted to approximately a 100-mile radius (160 km) around Sydney, Australia where they occur in humid forests. Males wander about seasonally searching for females. Females have a limited territory, thus wandering males inflict most bites. Several other funnel web spiders are known to inflict serious bites, but are not known to cause fatalities. Envenomation of humans by these spiders

Funnel web spider (*Atrax robustus*) Australia.
Source: Stephen L. Doggett

Distribution of funnel web spiders (*Hadronyche* spp.).

can produce serious medical consequences, including death. However, since the late 1920s, there have only been around a dozen deaths attributed to these spiders.

These moderate sized spiders (some approaching 38 mm or 1.5 inches long) have a formidable reputation of gripping their victim and repeatedly biting. Their neurotoxic venom is unusual in that it is only toxic to primates (monkeys and humans), who lack a naturally occurring inhibitor. Both male and female funnel web spiders are toxic, but males are more virulent even though females usually produce more venom.

Once injected, venom can reach the circulatory system in as little as 2 minutes. Envenomation reactions include skeletal muscle spasms and twitching, weakness, excessive salivation and sweating, bristling of hairs, rapid heartbeat, high blood pressure, irregular heartbeat, abdominal pain, nausea and vomiting, pulmonary oedema, capillary leaking, kidney failure, unconsciousness, shock and death. Death can take as little as 15 minutes or up to 3 days. If bitten by a funnel web spider, first aid measures should be applied and professional medical treatment should be obtained as soon as possible. Australia has antivenom available.

Mouse spiders
(Genus *Missulena*)

Mouse spiders are medium to large bodied (up to 3 cm or 1.2 inches) in size. Most mouse spiders occur in Australia (~16 species) and one additional species is known from Chile. They tend to be secretive and live in burrows so they are not commonly encountered. Although not aggressive, they will bite defensively and can inflict painful, often dry, bites without delivering venom. However, if venom is delivered the bites are cause for concern because the venom is thought to be similarly toxic to that of the funnel

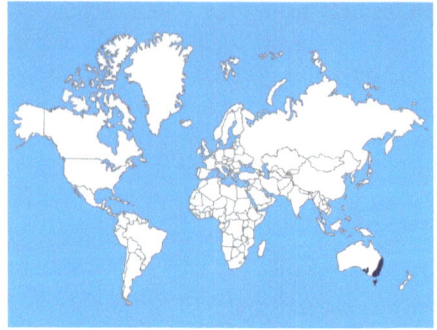

Distribution of funnel web spiders (*Atrax* spp.).

Mouse spider (*Missulena occatoria*). Source: Peripitus/Wikipedia. Cc BY-SA 3.0. Available from https://en.wikipedia.org/wiki/Missulena#/media/File:Male_Missulena_occatoria_spider_-_cropped.JPG

web spiders (indeed, some bites from mouse spiders have produced symptoms similar to those of the funnel web spiders). Funnel web spider antivenom has been successfully used to treat some victims of mouse spider bites. Because several species of mouse spiders have been implicated in human bites in Australia, the entire group must be considered potentially dangerous.

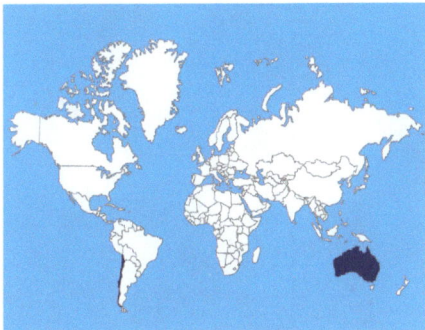

Distribution of mouse spiders (*Missulena* spp.).

Baboon spiders
(Genera *Harpactirella*, *Pelinobius*)

The baboon spiders are African tarantulas. They can exceed 32 mm (1.25) in length and have very long spinnerets protruding from the tip of the abdomen. The body and legs are covered thickly with brownish-grey hair-like setae. They reside in silk-lined tunnels beneath rocks and logs.

Although most baboon spiders do not pose a significant threat, their venom is strong and one species, *Harpactirella lightfooti*, can inflict severe bites. A bite from this spider may initially result in discolouration, swelling and a burning sensation at the bite site. After a 2-hour period, repeated vomiting may occur, in addition to symptoms of shock. The victim may then collapse, and be unable to walk. It is quite possible other species of baboon spiders produce similar bite responses. Laboratory testing with mice shows *Latrodectus* antivenom is a promising treatment. The exact distribution of *H. lightfooti* in Africa is not well established, but it appears restricted to the South-western Cape region. The king baboon spider (*Pelinobius*

King baboon spider (*Pelinobius muticus*). Source: www.universoaracnido.com/Wikipedia. CC BY-SA 2.5. Available from https://upload.wikimedia.org/wikipedia/commons/7/75/Pelinobius_muticus_adult.jpg

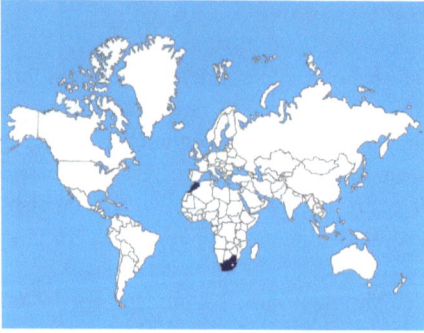

Distribution of baboon spiders (*Harpactirella* spp.).

Six-eyed sand spider (*Sicarius terrosus*). Source: Beliar spider/Wikipedia CC GFDL. Available from https://en.wikipedia.org/wiki/Sicarius_(spider)#/media/File:Sicarius_terrosus,_female_-_02.jpg

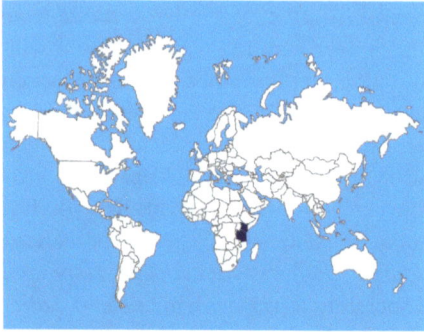

Distribution of king baboon spider (*Pelinobius muticus*).

muticas), of eastern Africa, is known for its aggressive, defensive behaviour. The bites from this large spider are reported to be very painful.

Six-eyed sand/crab spiders
(Genus *Sicarius*)

The 22 known *Sicarius* species inhabit Southern Africa, Central and South America, and the Galapagos Islands. They are medium-sized spiders up to 15 mm (0.6 inch) long, with about a ~50 mm (2 inches) wide leg span. Most are reddish-brown to yellow in colour without any distinct patterns. They often wedge sand particles between body hairs, thus camouflaging themselves against a similar background.

These spiders are shy, secretive and seldom encountered, but they will bite if accidentally contacted.

Six-eyed spiders are arguably the most venomous Southern African spider group. Their cytotoxic venom is a virulent toxin that destroys tissue around the bite and throughout the body. The venom can result in extensive internal bleeding and severe tissue and organ damage, and general distress. Severity depends on the location of the bite, amount of venom delivered, and the health of the victim, including allergies, and age. Small children and the elderly tend to be the most adversely affected. Experiments have shown that *Sicarius* venom injected into rabbits killed

Distribution of six-eyed sand spiders (*Sicarius* spp.).

them within 4 to 6 hours. Autopsies of the rabbits revealed extensive subdermal, skeletal muscle, heart and kidney tissue damage, plus swelling of the liver and blockages of the lung arteries. No antivenom is available for this genus. If bitten, whether known or suspected, the victim should seek immediate medical treatment following first aid treatment.

Wandering spiders
(Genus *Palystes*)

This is the largest South African spider with females reaching 41 mm (1.6 inches) long and the males being only slightly smaller. The various common names given to this group of spiders include wandering spider, rain spider, lizard-eating spider and dwaalspinaekop. Although there are ~12 species of *Palystes*, symptomatic bites are most often attributed to *P. superciliosus,* commonly known as the rain spider. Wandering spiders can be confused with the baboon spiders, but the eyes of wandering spiders are arranged as two sets of four, rather than a single, small clump like those of baboon spiders. Other distinguishing

Distribution of wandering spider (*Palystes superciliosus*).

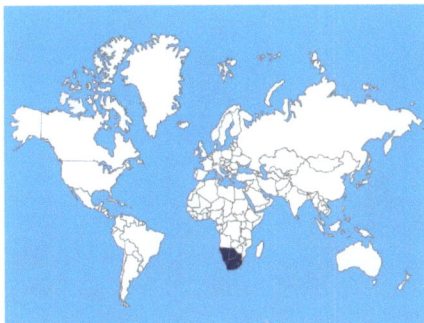

Rain spider (*Palystes superciliosus*). Source: JMK/ Wikipedia. CC BY-SA 3.0. Available from 184020033https://commons.wikimedia.org/wiki/ File:Palystes_superciliosus,_a,_Pretoria.jpg

characteristics of *P. natalius* are a brownish-grey coloured body, bright yellow ventral leg surfaces with transverse black bands, and a reddish oral region. These free-living spiders are often found running on the walls of houses, particularly in rural areas. *Palystes superciliosus* is included here as a medical threat because its venom has caused convulsions and death in guinea-pigs in medical experiments. However, some researchers argue the spider's large, piercing chelicerae killed the guinea-pigs due to shock rather than venom toxicity. One recorded human bite only produced burning pain and slight swelling that persisted for a few days.

White-tailed spiders
(*Lampona cylindrata, Lampona murina*)

White-tailed spiders grow to 20 mm (0.78 inch) long, and they typically inhabit homes and cool, outdoor locations under tree bark, rocks and leaf litter throughout Australia. The common white-tailed spider, *Lampona cylindrata*, has been introduced to New Zealand. Their bite can cause a burning pain followed by localized swelling and itching. Reports of bites from

White-tailed spider (*Lampona cylindrica*). Source: Stephen L. Doggett

Yellow sac spider (*Cheiracanthium inclusum*). Source: David E. Bowles

these spiders causing necrosis similar to that of a brown recluse have been largely debunked and research has shown the venom has little potential to cause necrosis. Rarely, reactions have been reported following white-tailed spider bites that are similar to necrosis, but these reactions are thought to be due to other factors such as contamination by bacteria (*Mycobacterium ulcerans*) on the spider's fangs or the skin, and not the venom.

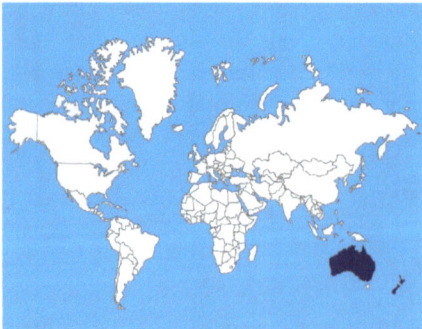

Distribution of White-tailed spiders (*Lampona* spp.).

Yellow sac spiders
(Genus *Cheiracanthium*)
More than 200 yellow sac spider species are distributed worldwide. They are relatively small bodied (length 10 mm or 0.4 inch), and, as the name implies, they are yellowish

in colour. They hide in foliage, under bark or stones where they construct sack-like, silken tubes. Although reclusive, they occasionally enter houses and other structures. Yellow sac spiders can be aggressive spiders and will bite defensively. Although their bite is considered painful, similar to a bee or wasp sting, it is not particularly dangerous. Localised redness, swelling and itching may follow the initial bite. Systemic effects usually are minimal and seldom reported. Symptoms typically dissipate entirely within a couple of days. Clusters of bites from *C. mildei* have been previously reported from some large cities in the United States. Similarly, *C. inclusum* is reportedly responsible for bites in the state of Georgia (US) and south-western Canada. Bites from *C. inclusum* are probably far more common and widespread than reported. First aid should be used as needed following bites, and if symptoms progress or worsen, professional medical help should be sought.

Woodlouse or wood hunter spiders
(Genus *Dysdera*)
Primarily European, the ~200 species of woodlouse spiders can also be found throughout the world. They resemble recluse spiders, but they lack the distinctive

Woodlouse spider (*Dysdera crocata*). Source: Hans Hillewaert_Wikipedia. CC BY-SA 4.0. Available from https://en.wikipedia.org/wiki/Woodlouse_spider#/media/File:Dysdera_crocata_(male).jpg

Tarantula (*Aphonopelma hentzi*). Source: David E. Bowles & Mark Pomerinke, United States Air Force

fiddle-shape on the cephalothorax. One European species, *Dysdera crocata*, is now widely introduced worldwide. Woodlouse spiders primarily prefer the outdoors, but they do occasionally enter houses and other structures. If threatened, they may become defensive and will readily bite. Bites may be painful and somewhat similar to a bee sting. Pain usually dissipates within a couple hours, although there may be localised residual itching for a few hours.

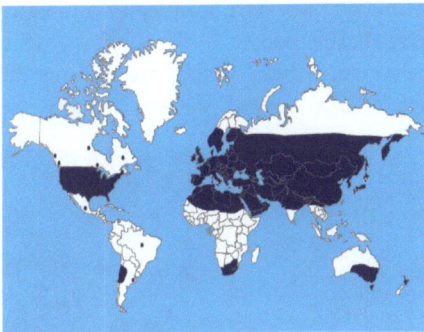

Distribution of woodlouse spider (*Dysdera crocata*).

Tarantulas
(Family Theraphosidae)
Though widely feared because of their large size and negative portrayals in the

media, tarantulas are not particularly dangerous. The several hundred known species of tarantulas are distributed worldwide. They have long, needle-like fangs and can inflict quite a painful bite, and they may shed urticating setae that can irritate eyes or mucous membranes. Their venom produces localised reactions comparable to bee and wasp stings. First aid measures usually suffice for bite victims.

Other potentially dangerous spiders
Other spiders implicated in arachnidism include *Argiope* spp. (garden spiders), found worldwide, and *Phidippus* spp. (jumping spiders) found primarily in the Western Hemisphere. Victims may exhibit localised pain due to the bites of these spiders and, although rare, necrotic envenomation reactions are generally mild. Similarly, some species of wolf spiders have cyanotic venom that may cause intense pain, localised necrosis, bite site reddening and variable swelling. Sometimes the powerful jaws of some wolf spiders may cause bleeding at the puncture sites. One South American species, *Lycosa raptoria*, produces necrotic lesions and victims may experience swollen lymph vessels around

Garden spider (*Argiope* sp.) North America. Source: David E. Bowles

Jumping spider (*Phidippus audax*). Source: Kaldari/ Wikipedia. CC0. Available from https://en.wikipedia. org/wiki/Phidippus_audax#/media/File:Kaldari_ Phidippus_audax_01.jpg

A wolf spider (*Hogna radiata*). Source: Papa Pic/ Flickr. Public domain. https://www.flickr.com/ photos/oscarfava/8676088114/

the bite area, with eventual eschar formation and wound sloughing. Treatment for most of these bites typically involves only first aid measures, but professional medical assistance should be obtained if symptoms persist.

Camel spiders
(Order Solifugae)

Camel spiders, windscorpions, sunscorpions or sunspiders, are solpugids and not true spiders. They are a distinct arachnid group containing ~900 species inhabiting Africa, Middle East, Mexico and southwestern United States. They are nocturnal animals that retreat to animal burrows or under rocks or other objects during the day, so they are not frequently encountered. They are not venomous, but their powerful jaws make them very efficient predators of arthropods, lizards, small birds and rodents. They can also inflict painful bites if mishandled, which accounts for most human bites. Bites can be treated with local analgesics if needed, but they are not a significant health threat. Although there have been reports of solpugids chasing people, it is thought that this behaviour is because they are seeking the shade of a person's shadow and not an

Camel spider (Solifugae). Source: David-O/Flickr. CC BY 2.0. Available from https://www.flickr.com/ photos/8106459@N07/36711739553/

A camel spider in defensive posture. Source: TodonFlickr/Flickr. CC BY 2.0. Available from https://www.flickr.com/photos/toddneville/3766176079/

attack of any sort. Solifugids are not large creatures as often depicted in social media with the largest known species being less than 15 cm (6 inches) in length, and most being much smaller.

Distribution of camel spiders (Solifugae).

Spider bite prevention and treatment

Situational awareness and effective personal protection measures are key to avoiding spider problems (see 'Personal protection measures'). Refer to the 'Education' section for recommended resources, and Appendix 6 to learn about the specific dangerous spiders where you are going. You should assess the relative spider threat, emergency

medical care locations and appropriate antivenom availability where you are going before travelling there. Remember, most spider bites are caused by accidentally or deliberately contacting them. Because repellents do not work on spiders, it is best to minimise risks of bites by using avoidance and proper wear of protective clothing, particularly gloves, during high spider bite risk activities (e.g. moving fire wood, cleaning up debris, moving boxes, stocking shelves, working in out-buildings, cleaning up neglected areas). Enclosed sleeping quarters are best (i.e. closed doors and windows, window/tent screens; sealed tent floors and zippered openings) because some spiders wander. Dwellings and other temporary or infrequently used dwellings or adventure equipment should be periodically inspected for spiders. This includes sleeping bags, tents, backpacks, luggage, hunting or fishing gear. Remove any spiders you find and seek professional pest management services if necessary.

Should you experience a mild spider bite reaction, thoroughly clean the immediate bite site, immobilise and elevate the affected area. To select the correct treatment strategy and particularly the correct antivenom (if necessary), it helps if the medical professional knows what spider bit you. Therefore, it is ideal if the offending spider can be captured in a sealed container or photographed to increase the chances of an accurate identification. Over the counter topical corticosteroids, systemic antihistamines and cold compresses can be used to treat localised reactions. Milder pain management can include aspirin and acetaminophen. However, any systemic symptoms or necrosis indicates a more serious situation and the victim should seek immediate professional medical treatment. Treatment varies depending on the severity of envenomation, and pain management may include using analgesics (possibly intravenous) or more substantial drugs such as diazepam, oxycodone and morphine. Aluminium sulfate compounds (Stingose®), antihistamines and corticosteroid injections can be used to reduce pain and inflammation associated with spider bites, and muscle relaxants (10% calcium gluconate intravenous solution) have been used to treat the neurotoxic effects of black widow venom. Dapsone (100 mg) is often used to limit the extent and severity of cutaneous necrosis, but surgical excision and skin grafts may be required for extreme necrosis although such cases are relatively rare.

Scorpions
(Order Scorpiones)

Over 1750 species of scorpion are widely distributed in tropical, subtropical and desert habitats worldwide, but generally south of 45°N latitude. They are largely nocturnal and secretive animals, and all are venomous. Most scorpions are not aggressive and they usually sting due to accidental contact. Most stings are similar to those inflicted by bees or wasps. However, a few species are dangerous and remain a serious public health menace in some areas. For example, scorpion stings are a leading cause of morbidity in Mexico, with roughly 250 000 stings and 100 deaths reported annually. Similarly, Tunisia had 30 000 to 45 000 stings and 35 to 105 deaths per year between 1986 and 1992, largely among children. Appendix 6 summarises the most dangerous scorpions and their distribution. Here we highlight a few of the most dangerous scorpions then cover prevention and treatment including first aid measures.

Dangerous scorpions

Most potentially lethal scorpions belong to the Family Buthidae, which primarily inhabit Africa, the Middle East, Asia and South America. Because accurate identification of scorpions can be difficult, all scorpions worldwide should be considered potentially dangerous. Many scorpions will sting, but cause only unpleasant, temporary pain with no lasting consequences. Several *Centruroides* species, which inhabit the United States, Mexico and southward, have serious medical consequences, while other *Centruroides* species only produce painful encounters.

Among the most dangerous scorpions with respect to human mortality are *Tityus serrulatus* (Brazil) and the infamous yellow scorpion, *Leiurus quinquestriatus* (Middle East). Several species of the scorpion genus *Hottentotta* have severely painful stings, and the Indian red scorpion, *Hottentotta tamulus*, is considered one of the most dangerous scorpions in the world. Historically, the human mortality rate attributed to the Indian red scorpion is around 30%. The Western Cape of Africa's most important venomous species is *Parabuthus granulatus*, while stings from the North African species *Androctonus australis* and *Buthus occitanus* regularly cause serious consequences. Another dangerous African species is *Parabuthus transvaalicus*, which can both sting and spray venom from its stinger (up to 1 m), and it is accordingly known as the spitting scorpion. A South African species, *Opistophthalmus glabrifrons*, can reportedly produce various dangerous systemic symptoms, but no deaths are known to have occurred. *Androctonus crassicauda* and *Buthus occitanus* are generally considered Jordan's most dangerous scorpions. A study of 2534 patients in south-west Iran showed three scorpion species caused nearly all stings: *Mesobuthus eupeus* (45%), *Androctonus crassicauda* (41%), and *Hemiscorpius lepturus* (13%), but the only the latter two species are considered dangerous. Indeed, *Androctonus*

Fattail scorpion (*Androctonus australis*). Source: Aaron Saguyod/Flickr. CC BY-ND 2.0. Available from https://www.flickr.com/photos/ajdcsaguyod/7109679613/

Fattail scorpion (*Androctonus crassicauda*). Source: Moshe Klukowski

crassicauda causes many deaths annually in Iran, mostly among children, and it is considered a significant medical hazard in that region. The yellow Iranian scorpion, *Odontobuthus doriae*, does not cause as many stings, but it has a potent and dangerous neurotoxic venom. The only scorpions occurring in North America capable of inflicting a fatal sting is the Arizona bark scorpion (*Centruroides sculpturatus*), which inhabits Arizona, California, Utah and north-western western Mexico, and *Centruroides suffusus*, found in Durango State, Mexico. However, the Arizona bark scorpion has not caused any deaths in the United States since 1968. The widely dis-

tributed striped bark scorpion (*Centruroides vittatus*) is often confused for the more dangerous Arizona bark scorpion, but it is capable of inflicting a painful sting without more significant consequences.

Most dangerous scorpions tend to have long, slender pedipalps (claws) and tails, while the less venomous species have thick, robust pedipalps and tails. The thought being those with smaller claws rely more on toxic venom to subdue their prey verses those who use their larger, more powerful claws for that purpose. However, do not assume thick-clawed scorpion stings are harmless or non-lethal. Within the known range of highly venomous scorpions, stings from any scorpion must be taken seriously and medical assistance obtained as soon as possible.

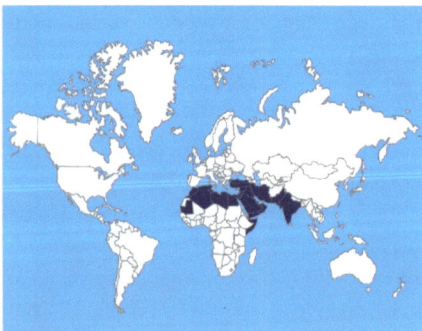

Distribution of fattail scorpion (*Androctonus australis*).

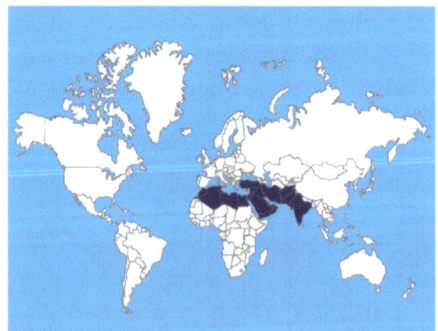

Distribution of fattail scorpion (*Androctonus crassicauda*).

Common yellow scorpion (*Buthus occitanus*). Source: Alvaro Rodríguez Alberich/Wikipedia. CC BY-SA 2.0. Available from https://en.wikipedia.org/wiki/Buthus_occitanus#/media/File:Buthus_occitanus.jpg

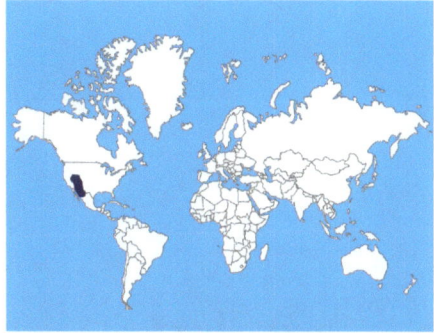

Distribution of Arizona bark scorpion (*Centruroides sculpturatus*).

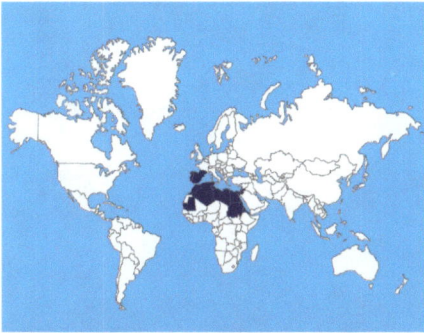

Distribution of common yellow scorpion (*Buthus occitanus*).

Stripped bark scorpion (*Centruroides vittatus*). Source: Wyatt Berka/Flickr. CC BY-SA 2.0. Available from https://www.flickr.com/photos/wyattberka/5907216873/

Arizona bark scorpion (*Centruroides sculpturatus*). Source: Matt Reinbold/Flickr. CC BY-SA 2.0. Available from https://www.flickr.com/photos/furryscalyman/293651238

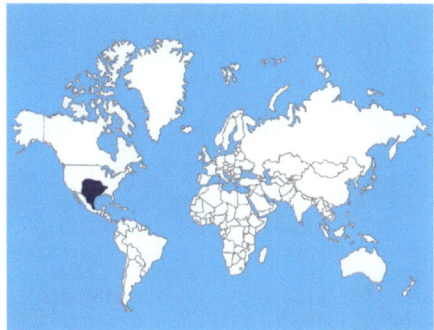

Distribution of stripped bark scorpion (*Centruroides vittatus*).

Distribution of *Hemiscorpius lepturus*.

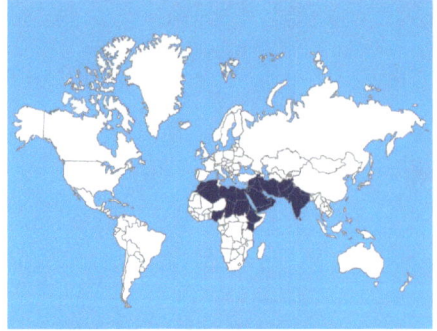

Distribution of *Hottentotta* spp.

Hottentotta jayakari. Source: Stefan Muth/Flickr. CC BY-SA 2.0. Available from https://www.flickr.com/photos/st_muth/8593058282/in/photolist-e6kERh-d7snu1-dbSRX3

Yellow scorpion (*Leiurus quinquestriatus*). Source: Alastair Rae/Flickr. CCBY-SA 2.0. Available from https://www.flickr.com/photos/merula/7169672743

Indian red scorpion (*Hottentotta tamulus*). Source: Shantanu Kuveskar/Wikipedia. BY-SA 4.0. Available from https://en.wikipedia.org/wiki/Hottentotta_tamulus#/media/File:Scorpion_Photograph_By_Shantanu_Kuveskar.jpg

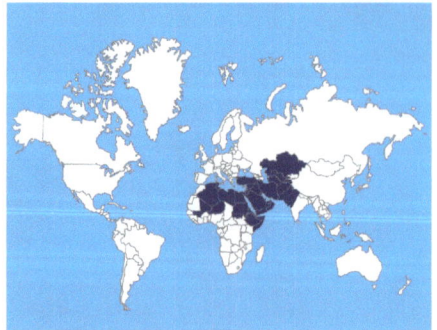

Distribution of yellow scorpion (*Leiurus quinquestriatus*).

Yellow Iranian scorpion (*Odontobuthus doriae*).
Source: Hectonichus/Wikipedia. CC BY-SA 3.0.
Available from https://commons.wikimedia.org/
wiki/File:Buthidae_-_Odontobuthus_doriae.JPG

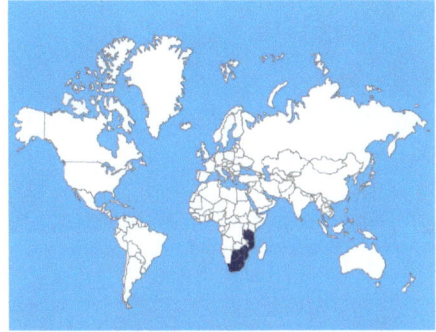

Distribution of Thick-clawed burrowing scorpion
(*Opistophthalmus glabrifrons*).

Distribution of yellow Iranian scorpion
(*Odontobuthus doriae*).

Granulated thick-tailed scorpion (*Parabuthus
granulatus*). Source: Bernard DuPont/Wikipedia
Available from https://commons.wikimedia.org/
wiki/File:Granulated_Thick-tailed_Scorpion_
(Parabuthus_granulatus)_%22_Brown_
Phase_%22_(7003255311).jpg

Thick-clawed burrowing scorpion (*Opistophthalmus
glabrifrons*). Source: Peter G.W. Jones/Flickr. CC
BY-ND 2.0. Available from https://www.flickr.com/
photos/flickpicpete/15202024372/

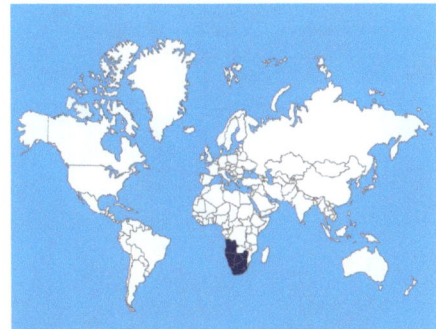

Distribution of granulated thick-tailed scorpion
(*Parabuthus granulatus*).

South African spitting scorpion (*Parabuthus transvaalicus*). Source: Bart Wursten

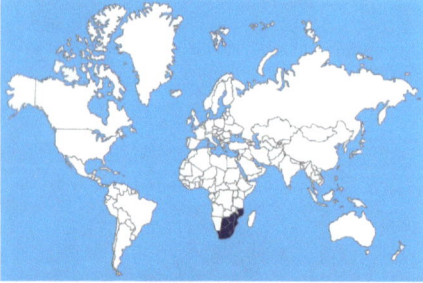

Distribution of South African spitting scorpion (*Parabuthus transvaalicus*).

Brazilian scorpion *(Tityus stigmurus)*. Source: Hjalmar Turesson/Wikipedia. CC0 1.0 Universal. Available from https://en.wikipedia.org/wiki/Tityus_stigmurus#/media/File:Tityus_stigmurus_20140919.JPG

Distribution of Brazilian scorpion (*Tityus stigmurus*).

Scorpion venom effects

Scorpion venom produces local and systemic effects that are often quite variable, ranging from self-resolving pain to death. Severity is species dependent and varies widely so it is highly preferred to capture or photograph the scorpion responsible for a sting for accurate identification and rendering of specific medical aid. Additionally, physiological responses vary depending on a person's general health, age, physiology, genetics and emotional state. Other influential factors include sting location, penetration depth, amount of venom injected, and whether or not the venom entered the circulatory system. Children and small-bodied adults tend to get a higher venom concentration per blood volume compared with larger bodied people, which often complicates the severity of the reaction.

Scorpion venom contains both haemolytic and neurotoxic components, the former producing the pain and swelling following stings. The reaction to a sting varies according to the species of scorpion, and reactions may vary widely and can present in complex combinations. Some

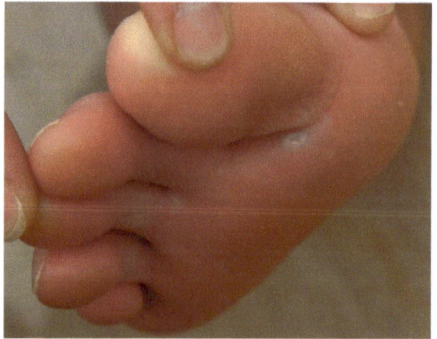

Blistered big toe caused by an unidentified scorpion sting in the Middle East. Victim only suffered initial pain, swelling and headache. Source: James A. Swaby

stings produce severe swelling and discolouration at the sting site, but no pain. Others cause pronounced swelling, inflammation and pain. Although the initial sting reaction is generally an immediate, localised burning pain, some potentially lethal species often cause little or no initial pain, swelling, inflammation and discolouration. However, the sting site may become painful to the touch and have a 'woody' feeling. In many cases, the victim develops dark blue skin patches, usually with a red halo, near the sting location within the first hour. These gradually become hardened, inflamed, and eventually necrotic with subsequent skin sloughing. Large blisters may also develop around the sting site followed by extensive ulceration. In severe envenomations, the haemolytic components of the venom destroy red blood cells, disrupt blood clotting, and cause other cardiovascular complications. Acute kidney failure, typically characterised by bloody urine, anaemia and jaundice from destroyed blood cells, can result within 24 hours to a few days after the sting, and may require kidney dialysis. When the kidneys are damaged, victims may slowly begin secreting abnormal amounts of urine 6 to 21 days after a sting.

Depending on species involved and quantity injected, the neurotoxic fraction can produce a broad range of dangerous and potentially fatal reactions. The various polypeptides making up this fraction disrupt the nervous system's ionic balance and ion-channel activity. The primary and initial effects are on the peripheral nervous system causing intense pain, altered heart activity and numbness. Other neurotoxic symptoms include muscle twitching, crying, salivation, profuse sweating, respiratory distress, urinary urgency, nausea, tongue paresthesia, restlessness, joint stiffness, convulsions and increased muscle activity around the eyes. Because some scorpion species only cause negligible pain initially, the victim may only experience gradually increasing local pain, swelling and inflammation before having to seek medical treatment hours to days later. Extreme restlessness characterised by excessive neuromuscular activity (jerking and spasms) is another common response children may show. Typically, blood pressure, body temperature and tendon reflexes increase, while motor skills become impaired. Other striking features caused by scorpion neurotoxins include the inability to write or manipulate small objects, difficulty articulating speech and variable pharyngeal reflex loss. Victims may also experience heightened muscle pain, cramps and sensitivity to touch, cold or heat. Potentially fatal systemic symptoms and signs usually develop within 4 hours. Anaphylaxis plus cardiac or respiratory failure have been shown to cause death within 24 hours. Respiratory dysfunction often complicates recovery to varying degrees, which tends to be more serious in children. Deaths are less common now due to the widespread availability of antivenom in areas where dangerous species occur. Travellers should always inquire about the availability of scorpion antivenom before travelling to areas with known occurrence of dangerous scorpions.

Scorpion sting prevention and treatment

Situational awareness and effective personal protection measures are the keys to avoiding scorpion stings (see 'Personal protection measures'). Plan ahead and refer to the 'Education' section recommended resources, and Appendix 6 to learn about the scorpions that occur in the area where you plan to travel. Because risks among scorpions are so variable, you should assess the relative threat, emergency medical care locations and availability of appropriate antivenom where you are going. Repellents do not work on scorpions, so wear protective clothing, particularly gloves, during high scorpion risk activities (e.g. moving firewood, cleaning up debris, moving boxes, stocking shelves, working in out-buildings, cleaning up neglected areas). In places where scorpions are found, be sure to inspect cabins and other temporary or infrequently used dwellings or adventure equipment before use. This includes sleeping bags, tents, backpacks, luggage, and hunting or fishing gear. Enclosed sleeping quarters are best (i.e. closed doors and windows, window/ tent screens; sealed tent floors and zippered openings) for minimising exposure to scorpions. Consult pest management professionals if scorpions become a persistent nuisance

Following a sting, collect the scorpion in a sealed container, if possible, or take a photograph so that an accurate identification can be made. Specific identification of the offending scorpion can be very useful for establishing the appropriated treatment strategy. First aid measures for scorpion stings include immobilisation, elevation of stung limbs and cold compresses. Victims should seek immediate medical attention for any symptoms beyond general pain. This is especially important for young children.

Medical professionals may make observations of the victim for at least 24 hours, particularly if a known dangerous species is involved, and this may involve hospitalisation. Treatments can range from applying a cold compress or ice pack on the sting site to use of antivenom. Medical practitioners also may use a local anaesthetic (e.g. xylocaine) or ice pack to relieve localised pain, but cases involving neurological symptoms may require medications such as barbiturates, diazepam and atropine to manage. Some analgesics such as morphine, demerol, codeine or other morphine derivatives, paraldehyde, valium and thorazine may actually increase venom toxicity, so providers typically administer those drugs cautiously. Corticosteroids, hydration, blood transfusion and diuretics sometimes are used to help manage severe cases. Antivenom is normally used only in cases of severe systemic symptoms. Success of antivenom therapy depends on application conditions (e.g. dose, route and injection time since envenomation) and/or antivenom quality.

Mites and chiggers (Subclass Acari)

Estimates suggest there are between 30 000 to 49 000 known species. They are found worldwide in a variety of habitats from terrestrial to aquatic. Most mites are microscopic, free-living scavengers or parasites. Some are parasites of animals that occasionally attack people causing irritation and potentially transmitting disease agents. Some of the more commonly encountered mites that sometimes parasitise people are treated here. Attacks by mites in some areas of the world may be caused by species different from those indicated here.

Chiggers or harvest mites
(Family Trombiculidae)

Most medically important chiggers belong to the genera *Eutrombicula* and *Leptotrombidium*, each with multiple species. The distribution for the family is broad and nearly global. Only the Saharan Desert and northern boreal areas lack these mites. However, populations tend to be greatest in tropical to warm temperate regions. Chiggers in the Genus *Eutrombicula*, which inhabit the Western Hemisphere and Europe, do not transmit any known human pathogens, but they can cause irritating bites, dermatitis and severe itching. By comparison, chiggers in the genus *Leptotrombidium* transmit scrub typhus in Asia and portions of Australia. In contrast to the bites from *Eutrombicula* mites, those of *Leptotrombidium* typically do not itch, or less intensely so, and a black, crusty, necrotic lesion called an eschar often forms at the bite site.

Chiggers are obligate ectoparasites on vertebrate hosts before moulting to the nymph and adult stages. Adult and nymphs are free-living and eat small invertebrates, their eggs and organic matter. Magnification is necessary to see the microscopic chigger, which is the larval stage. In contrast, the bright red, eight-legged adult, or harvest mite, is readily visible to the naked eye. Chiggers are most likely encountered in habitats where their rodent hosts are found, including brushy or grassy areas. Females may lay several hundred eggs on the ground in several groups, and these clumps hatch and form larval clusters. When an animal, including a person, walks through or brushes against infested vegetation the chiggers actively crawl onto the body. Once on, the larvae move to an ideal feeding spot and tightly attach themselves to your skin. Contrary to popular belief, chiggers do not burrow into the skin or suck blood. They pierce skin (often around a hair follicle) to feed on pre-digested host tissue and lymph. To feed, chiggers introduce digestive enzymes that liquefy surrounding tissues. The enzymes inflame the tissue around the bite forming a red welt. They also extrude a proteinaceous salivary material that quickly hardens into a ring. They then ingest some of the dissolved host tissue from the middle of that ring, secrete more saliva and then another hardening ring.

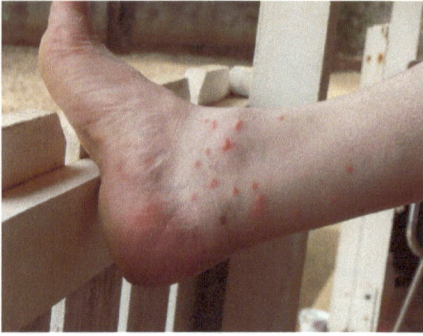

Chigger bites on lower leg. Source: Elin/Flickr. CC BY 2.0. Available from https://www.flickr.com/photos/beckmann/436378641/

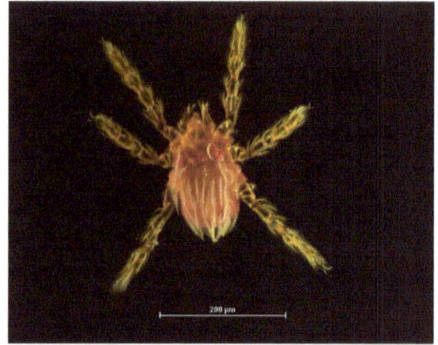

A chigger (*Leptotrombidium* sp.). Source: Piyada Linsuwanon and Nutthanun Auysawasdi, Department of Entomology, AFRIMS, United States Army

This is repeated many times until the chigger has fully engorged or has been physically dislodged from the host's skin surface. The stack of successive hardened rings form a minute, pale whitish tube (stylostome) that remains in the middle of the bite site after the chigger has departed. This rash and associated intense itching are an allergic reaction to the mite's salivary secretions. Itching may last for several days, but is generally most intense during the first 2 to 3 days.

When full, the chigger drops off the victim, goes into organic debris on the ground and enters a quiescent stage until it moults into a predatory nymph stage that feeds and moults, finally becoming the overwintering bright red adult. Chigger mites produce one generation each year, and are most abundant during late summer and early autumn. Host seeking behaviour of larvae drops to zero following the first hard freeze.

Hot showers/baths, over-the-counter topical corticosteroids, antihistamines and

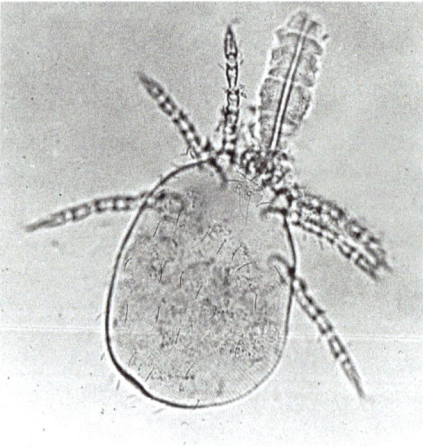

Chigger (*Eutrombicula* sp.). Source: Alan R. Walker/Wikipedia. CC BY-SA-3.0. Available from https://en.wikipedia.org/wiki/Trombiculidae#/media/File:Trombicula-mite-larva-with_stylostome-2.jpg

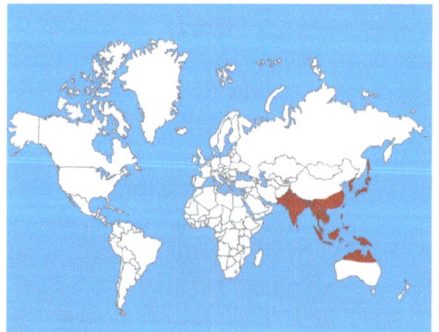

Distribution of scrub typhus.

aluminium sulfate compounds (Stingose®), can help alleviate the itching. Applying rubbing alcohol to the bites also serves to minimise itching. Try to avoid scratching to prevent secondary infections. People who engage in outdoor summertime activities in areas where chiggers are found should take a hot bath or shower as soon as possible after coming indoors to help minimise chigger bites. Severe dermatitis may require professional medical treatment.

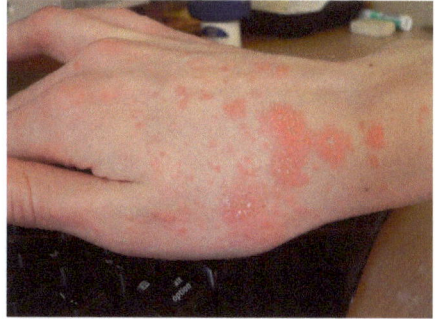
Scabies rash on hand. Source: Wikipedia

Scabies mites

Unlike chiggers, scabies mites (*Sarcoptes scabei*) are obligate parasites that feed solely on humans. The life cycle is rapid, lasting 14 days or less and all life stages, except eggs, are parasitic. The mites are tiny <0.4 mm (<0.01 inch) long, so positive identification requires skin scrapings and microscopic identification by trained professionals. They mainly spread person to person via direct contact or by exchanging infested clothing.

When scabies mites attack the host, they either produce burrows in the skin or reddened rash-like lesions. The female mite creates either intact (within) or exco-

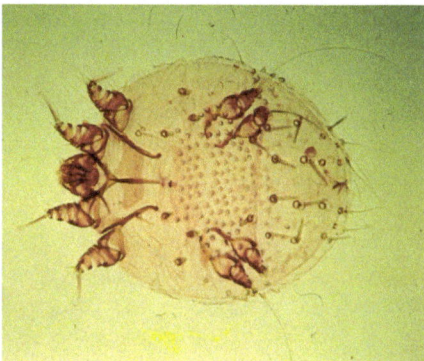
Scabies mite (*Sarcoptes scabei*). Source: J. Cross, USA Department of Defense

riated (open to the surface) burrows as she tunnels into human skin while laying eggs. The intact burrows are distinctly raised, linear, and with reddened marks, while excoriated burrows can become secondarily infected to form pustules and encrustations. Most infestations are on the extremities (i.e. hands, wrists, elbows, armpits, breasts and genitalia), but some are widespread over the body. The reddened rash-like lesions are most commonly found on the trunk.

Various medically prescribed topical medications, most commonly containing permethrin, are successfully used to treat scabies, but itching may continue several weeks during treatment. Topical corticosteroids, systemic antihistamines and antipruritic (itching) ointments can reduce itching. Scratching may result in secondary infections and should be avoided. Scabies-contaminated bed linens and clothing should be washed in hot water to minimise re-infestation.

Dust mites
(Genus Dermatophagoides*)*

The American house dust mite (*Dermatophagoides farinae*) and the European house dust mite (*D. pteronyssinus*) may

European house dust mites (*Dermatophagoides pteronyssinus*). Source: Gilles San Martin/Wikipedia. CC BY-SA 2.0. Available from https://en.wikipedia.org/wiki/House_dust_mite#/media/File:House_dust_mites_(5247996458).jpg

affect human health. Both mites are likely distributed worldwide. Their microscopic size, <300 μm (<0.01 inch) long, and translucent bodies make dust mites invisible to the naked eye. They live in bedding materials, furniture, carpet, stuffed toys and old clothing.

Dust mites feed on dead skin (dander) of humans and animals and other organic material. Some people experience allergic reactions to dust mites and their faecal pellets. They have also been implicated as a primary cause of many asthma cases. Symptoms may include sneezing, itching, watery eyes and wheezing. Occasionally a red rash develops on the skin, especially around the neck. Other allergic reactions may include headaches, fatigue and depression if infestations are long lived. Beds and mattresses may harbour millions, and are typically the most heavily infested items in houses. Carpeting and upholstery can also support large populations. Dust mites thrive in warm, moist environments. Although complete control of dust mites is not possible, the most practical approach to reducing populations and associated allergens is source reduction. This can be best accomplished through sanitation and using a dehumidifier to maintain relative humidity below 50%. Additionally, washing bedding materials, including pillowcases, sheets, blankets and mattress pads every other week in hot water (54°C or 130°F) or using enclosed mattresses, box springs and zipped allergen- and dust-proof covered pillows helps reduce dust mites. Highly sensitive individuals may require extensive mitigation measures such as frequent carpet cleaning and dusting, followed by immediate disposal of the dust bag. Alternately, replacing carpet with tile or wooden floors may resolve issues with dust mites. Eliminating or reducing fabric wall hangings such as tapestries or pennants, and covering or replacing upholstered furniture is also beneficial. Using HEPA type filters in air conditioner or heater vents are not practical or necessary. There use may actually aggravate mite problems because the small holes force air out at higher velocities thus stirring up more dust. Similarly, chemical control has no lasting effect on dust mite populations.

Other medically important biting mites

Although relatively rare compared with chiggers and scabies mites, several mites that normally infest various rodents, birds and insects occasionally attack humans when their hosts die or abandon their nests or other harbourage. These semi-transparent mites are not easy to see until they feed and become reddish to blackish coloured. Attacks are often related to occupational exposure, and self-limiting once you are no longer exposed. These mites attach only long enough to take a blood meal and then leave the host.

Unlike chigger or scabies mites, these mites are blood feeders and the initial bites can be mildly painful. Depending on the physiological condition of the victim and the number of bites inflicted, reactions to bites can be localised or widespread over the body. Toxic secretions produced by the mite during feeding further compound the allergic response to the bite. Reddish papule lesions with haemorrhagic centres may appear around the puncture wounds, or occasionally a fluid-filled vesicle appears. Lesions itch intensely, and may become crusted and secondarily infected. Bites are easily mistaken for other arthropod bites. Many people also report an irritating sensation on the skin when the mites crawl about. Topical corticosteroids, systemic antihistamines and anti-pruritic ointments such as crotamiton (e.g. EURAX®) can greatly reduce itching. Rubbing alcohol applied directly to the bites can also minimise itching. Scratching should be avoided to prevent secondary infections. Avoiding habitats where mites may occur and using personal protection measures are the best means of avoiding these mites, but, if exposed, shower and launder your clothing as soon as possible to reduce bites.

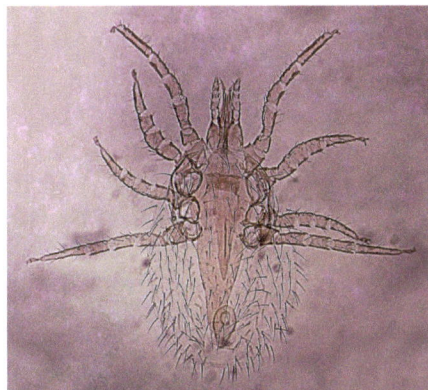

Tropical rat mite (*Ornithonyssus bacoti*). Source: Stephen L. Doggett

House mouse, tropical rat and spiny rat mites

The house mouse mite (*Liponyssoides sanquineus*), tropical rat mite (*Ornithonyssus bacoti*) and spiny rat mite (*Laelaps echidnina*) feed on rodent blood, but occasionally attack humans causing painful bites. They normally feed at night and hide in cracks or crevices around rat nests in buildings during the day. When rodents die or populations are very large, the mites leave their hosts and congregate around heat sources such as hot pipes and stoves. This behaviour increases the risk of people becoming the replacement hosts. House mouse mites occur throughout North America, northern Europe, Asia and Africa. The tropical rat mite is not restricted to the tropics, but it also inhabits temperate regions worldwide. The spiny rat mite occurs worldwide. House mouse mites transmit rickettsial pox (*Rickettsia akari*) that can sicken people with flu-like symptoms, but tropical rat mites and spiny rat mites do not transmit any known human diseases.

Bird mites

These mites sometimes incorrectly called 'bird lice' widely inhabit temperate and tropical regions worldwide. They occasionally attack humans causing painful, irritating bites, but they do not transmit any known human diseases. Infestations occur during early spring or summer when birds enter building attics and ceilings through gaps to construct their nests or roost outside on ledges or awnings. Avoid these areas, but, if exposed, immediately shower and launder your clothing to minimise risk of bites. Bird mites only live ~7 days, but they rapidly reproduce, generating large numbers of offspring, which disperse to attack humans after the young birds leave the nest or die. They typically feed at night and hide in cracks or crevices during the day. The most commonly encountered bird mites are the tropical fowl mite or starling mite (*Ornithonyssus bursa*), northern fowl mite (*Ornithonyssus sylviarum*) and red poultry or chicken mite (*Dermanyssus gallinae*), which primarily infest chickens and wild birds.

Straw itch and oak leaf itch mites

Straw itch mite (*Pyemotes tritici*) and oak leaf itch mites (*Pyemotes herfsi*) are insect predators that occasionally attack humans. Their bites are irritating and similar to those of chiggers, but they do not transmit any known human diseases. The oak leaf itch mite, native to Europe, also inhabits Australia, India, Egypt, Chile and many areas of the United States. Their populations are often centred near oak trees, hence their common name. The first reported cases of human bites in the United States were from Kansas in 2004, but outbreaks have since occurred in Illinois, Nebraska, Ohio, Oklahoma, Missouri, Pennsylvania, Tennessee and Texas. The females primarily feed on insects such as the oak marginal leaf fold gall larvae (a midge, Family Cecidiomyiidae), but they also attack stored product insects and even bark beetles. Each generation only takes about a week to develop, so populations explode over the summer and large numbers fall to the ground during late summer/fall to overwinter. Most people are

bitten when they encounter these mites outside, but, because they are easily carried by wind, they can be found indoors as well. During outbreaks, the best means of minimising bites is to use personal protection measures while outdoors and immediately shower and launder your clothing upon returning inside.

Straw itch mites ('hay' or 'grain itch mite'), found worldwide, heavily infest stored food products including fruits, seeds, cereals, pet food, grains, dried beans and peas. They also infest straw, hay, other dried grasses, and landscaping materials such as mulch, pine straw and wood chips under warm, humid conditions. The mites prey on insects that are found in these materials. People can get thousands of bites handling mite-infested landscaping materials, straw, crops such as beans, cotton, and small grains, crop residues, and similar materials, or while making

Straw itch mite (*Pyemotes tritici*). Source: Gary Bauchan and Ron Ochoa, United States Department of Agriculture, Agricultural Research Services, Electron & Confocal Microscopy Unit

dried plant arrangements. These mites can be carried via air currents and they can be blown indoors. Besides causing intense itching, severe cases can cause infection, fever, vomiting and joint pain. To reduce exposure, you should avoid warm, humid pantries, attics, crawl spaces, insulation, wall voids, warehouses, barns, grassy areas, weedy areas and landscaping materials, and use personal protection measures if these cannot be avoided. Immediately showering and laundering clothing helps reduce bites following exposure.

Cheyletiella mites ('rabbit fur mite')

Several species of these mites are distributed worldwide, infesting various birds and mammals, including cats, dogs and rabbits. They also prey on other mites and insects. They can cause a mange-like condition on pets, and they can be easily contacted when handling such infested animals. Their microscopic eggs cling to bedding, clothing, toys, rugs, furniture, and so on, which can make infestations difficult to eliminate. Although they do not reproduce while on people, they can rapidly reproduce on your pets, and they may live off any host up to 10 days. Such behaviour means they can result in serious infestations that are difficult to control. Feeding by these mites may cause an itchy rash condition termed *Cheyletiella* dermatitis or 'walking dandruff', because the mites carry around sloughed skin scales as they move over the host. The best means of avoiding these mites is to avoid or minimise contact with infested pets. Infected pets and homes must be treated promptly to halt the infestation. Attacks typically resolve once pet infestations are dealt with. Self-treatment can include

Rabbit fur mite (*Cheyletiella parasitivorax*). Source: Harold J. Harlan

over-the-counter topical corticosteroids, systemic antihistamines and anti-pruritic ointments. Medical assistance may be needed for severe skin rashes.

Flour and grain mites (Genus Acarus)

These mites (*Acarus siro*) are important pests of grain, dried fruit and vegetables. Grain mites are widely distributed throughout temperate regions worldwide and less commonly in tropical regions. They are tiny, 0.8 mm (<0.03 inch) long, pale, soft-bodied, pearly or greyish-white mites, with variously coloured, pale yellow to reddish-brown legs. Each leg has a single claw, and the males have enlarged forelegs with a thick spine on the bottom side. These two characters are used to separate *Acarus* from other mite genera although such determination must be made be trained professionals. Crushed mites give off a 'minty' odour, and when they occur in high numbers, this odour can be distinctive. Grain mites thrive under high moisture conditions and are often associated with fungal growth. Severe infestations of grain can produce 'mite dust': a brownish tinge that occurs on the surface of the grain. Although these mites do not feed on people, they do cause an itchy rash termed 'grocer's itch'. Sensitive individuals exposed to the shed setae and spines of these mites may develop allergic reactions. Infested food products are best avoided to reduce the potential for allergic reactions. Highly sensitive individuals should seek professional medical treatment.

Ticks
(Subclass Acari)

As disease vectors, ticks are second only to mosquitoes globally, and they are the most important disease vectors in North America. There are over 896 tick species or subspecies worldwide in over 18 genera; ~100 are medically important. Ticks are free-living external parasites of wild animals (birds, mammals, reptiles) and they require one or more blood meals for survival and reproduction. Their need for blood, persistence in host seeking, small size, cryptic behaviour, mostly painless bites, wide host range, longevity, relatively tough bodies and high reproductive potential (20–20 000 eggs per female) make them formidable biting pests and disease vectors.

The two medically important tick families are the Ixodidae (hard ticks) and the Argasidae (soft ticks). A third family, Nuttalliellidae, contains a single South African species that does not bite humans. Hard ticks are more diverse, more abundant and transmit most tick-borne human diseases and afflictions. In comparison, soft ticks are less diverse and less frequently encountered, and they are the primary vectors of only relapsing fever and Q fever.

Medical importance and disease transmission

Ticks and the diseases they transmit are complex subjects involving many tick species, disease agents, and biologies. It is a challenge to distil this vast amount of information to a level appropriate for this guide. Therefore, we briefly present their general biology and life history, the two major groups of ticks and a few representatives each, and summarise their overall medical importance. We also cover prevention of tick bites and treatment of the major diseases. To assist you in assessing risk from ticks and tick-borne disease, refer to Appendices 6 and 7, which list common medically important ticks and their general distributions, and summarises common tick-borne diseases and their tick vectors. The section on 'Personal protection measures' describes how to protect yourself from ticks, and Appendix 6 outlines proper tick removal procedures. Unlike many invertebrate threats that can best be described as irritants, several tick-borne diseases can be fatal or result in catastrophic disease. Therefore, we urge travellers to understand the relative risk from ticks when and where they travel. We recommend getting up-to-date status on the ticks, tick-borne disease risks, available vaccines, travel advisories and vaccination requirements before you travel. This information will give the reader an informed confidence when ticks are encountered.

Localised tick bite reactions

Localised tick bite reactions include swelling, erythema, paresthesia, blistering, itching, discolouration, skin hardening, and necrosis (dermatosis). In addition, improper tick removal procedures can leave mouthparts imbedded, leading to nodule formation. Without wound disinfection, secondary infections and localised gangrene can occur. Although uncommon, systemic symptoms can also occur including nausea, fever, vomiting, diarrhoea, irregular pulse, shortness of breath, gastrointestinal irregularities, restlessness, muscular weakness, drooping eyelids, light sensitivity, delirium, hallucinations and generalised pain. However, tick-borne disease and tick paralysis symptoms can overlap these symptoms, making diagnosis and treatment difficult for medical professionals.

Tick paralysis

Tick paralysis occurs worldwide and can be a very serious condition, resembling poliomyelitis. A salivary neurotoxin is thought to cause the paralysis. Certain ticks probably secrete toxin every time they feed but paralysis requires prolonged tick attachment (5–7 days). Long hair often conceals the stealthy ticks, particularly at the nape of the neck, especially among girls. Children typically are more vulnerable to this affliction than are adults. Routine body checks are recommended following adventures in tick habitats. Initial symptoms of tick paralysis onset include becoming restless and irritable, along with experiencing numbness or tingling of extremities, lips, throat and face. Weakness in the lower extremities follows the initial symptoms, which migrates to the trunk musculature, upper extremities, and head within hours or days. This causes difficulty walking and inability to stand. Eventually symptoms can progress to complete paralysis of the extremities, difficulty swallowing, slurred speech, double vision, coordination problems, respiratory failure and death (up

to 10% of cases). Over 40 tick species have been implicated in causing tick paralysis worldwide but *D. andersoni, D. variabilis, I. ricinus, I. holocyclus* and *R. evertsi* are most commonly involved. Correct removal of the offending tick usually leads to rapid and complete recovery (see Appendix 6).

Tick-bite alopecia

Another unusual condition stemming from ticks is tick-bite alopecia, which is hair loss around the bite wound. A reaction to salivary toxins produced by the tick is the likely reason for this response. Hair loss patches up to 50 mm (2 inches) and necrotic scarring may result. Tick-bite alopecia is self-limiting and hair regrowth is usually complete within 2 months, although scarring may be long term.

Otoacariasis or parasitic otitis

Otoacariasis or parasitic otitis is another bizarre and very rare response to ticks. When certain ticks enter the ear canal, they can cause a very irritating and often painful infection. Left untreated, the ticks can damage the eardrum or even the middle ear. However, it is highly unlikely you will tolerate the symptoms that long. Nearly all cases of otoacariasis are attributed to the spinose ear-tick, *Otobius megnini*, a livestock pest that has a worldwide distribution

Tick-borne diseases

Tick-borne diseases occur worldwide and normally circulate among ticks, livestock and/or wildlife, but they also occasionally infect humans. Ticks carry and transmit over 35 bacterial and more than 20 viral diseases. Globally, tick-borne disease rates pale compared with many other arthropod-borne diseases such as malaria, dengue, Chagas' disease, onchocerciasis and leishmaniasis, but tick encounters and tick-borne diseases have dramatically increased over the past 30 years worldwide. They are now the most widespread and medically important vector-borne diseases in Europe and the United States. Lyme disease (Lyme borreliosis) is now the most common vector-borne disease in the United States, with over 25 000 cases diagnosed annually, and over 65 000 cases are reported annually from throughout Europe. Appendix 7 lists over 55 tick-borne diseases, summarises the pathogens, onset symptoms, advanced symptoms, severity, distributions and tick vectors.

The surge in tick-borne diseases over the past 30 years is due to a complex set of interacting factors that includes changing land use patterns, climate change that alters natural tick host (deer, rodents, birds, etc.) populations, and associated tick population growth. Ecological and social upheavals, frequent international travel and international commerce all serve to promote tick dispersion and human–tick interaction. Tick-related factors influencing tick-borne infection rates include their prevalence, host prevalence and readiness to feed on people, which is compounded by multiple blood-feeding stages, multiple blood meals and multiple hosts that collectively serve to promote disease pathogen transmission. Some ticks also pass pathogens (i.e. spotted fever, recurrent fever and tularemia – see discussion under 'Fleas') from one generation to the next (transovarial transmission) and/or one life stage to the next (transstadial transmission). The primary physical mechanisms of pathogen transmission by ticks are varied. They include: (1) injected salivary fluids (i.e. spotted fever group rickettsiae, Lyme disease and tick-borne relapsing fever); (2) regurgitated midgut contents (i.e. Lyme disease); (3) infected faeces (i.e. Q fever); and (4) crushing a tick and transferring pathogens through abraded skin or the eyes (i.e. rickettsioses, ehrlichioses, Lyme disease, tick-borne relapsing fever, tularemia and Q fever). Others ticks, especially soft ticks, may secrete coxal fluids that contain pathogens (i.e. tick-borne relapsing fever), which can facilitate transfer to people who touch the infected ticks. To confound this threat, a single tick can carry two to three different pathogens. Thus, a single tick can potentially infect a host (person) with more than one disease. For example, in the United States, a single *I. scapularis* may carry Lyme disease (*Borrelia burgdorferi*), human granulocytic ehrlichiosis (*Ehrlichia equi*) and babesiosis (*Babesia microti*). Such multiple infections are extremely rare, but can cause atypical symptoms, making differential diagnosis difficult.

Symptoms of tick-borne disease infection vary, but onset is usually abrupt and flu-like (fever, headache, malaise, etc.). Some cause a rash on the trunk or the extremities, and some have multiple remission and illness cycles (see Appendix 7). However, tick-borne infections can mimic many other infectious diseases, including malaria, dengue, leptospirosis, hanta virus and meningococcemia. Fortunately, most tick-borne infections are mild, but without treatment, they can develop into serious, potentially life-threatening illness. Many such illnesses are easily treated with antibiotics, while others, such as viruses, do not respond to antibiotics and are quite difficult to treat. A few disease-specific vaccines are available should your travel objective require you to go to very high-risk areas. Travellers should consult with medical professionals on vaccine availability before they travel to high-risk areas. The guidance provided in Appendix 7 and on the website of the World Health Organization, provides information on the geographic areas of greatest concern.

Soft ticks
(Family Argasidae)

There are ~177 argasid species in four genera, but only three genera (*Argas, Otobius* and *Ornithodoros*) contain species of medical consequence to people. Soft ticks are found worldwide, but they are primarily distributed in tropical or subtropical regions. However, some species occur in desert or semi-desert areas, surviving very dry conditions and others prefer dry steppes and savannas.

Unlike the hard ticks, argasids have a more flexible body thus the name 'soft tick.' They tend to have focal habitats and typically spend their entire life in or near the habitats of their natural hosts. Typical habitats include mammal burrows, lizard shelters, bird rookeries, chicken houses, pigeon lofts, bat caves, stables and watering sites. Argasids have two to eight instars (normally three to four); usually all but the first require one complete blood meal to moult. A few larval argasids do not require a blood meal to moult. All adult argasid ticks require a blood meal to produce viable eggs. Females feed and lay eggs several times. Some soft ticks may complete five moults to complete their life cycle, feeding on a new host each time. During its lifetime, each soft tick takes five to 12 or more blood meals. They can also go many

years without a blood meal. Unfed soft ticks readily attach to a person who enters their habitat. They crawl to the host, quickly puncture the skin with their mouthparts, and rapidly engorge. Some soft tick bites are extremely painful, some even venomous, while others are painless. The bite of *Ornithodorus coriaceus* is very painful and may result in long-lasting effects. Other species are considered nuisance pests such as the Persian fowltick, *Argas persicus,* which readily bites people, particularly at night. The spinose ear-tick, *Otobius megnini,* occasionally invades the ear canal of its host causing severe pain called otoacariasis. After feeding, the ticks immediately drop off the host to seek hiding places among crevices, loose soil or cave walls to digest their meal, and develop or lay eggs, depending on the life stage. Attachment briefly lasts from about 10 minutes to a few hours. Attacks often occur among campers while they are sleeping in infested habitats. Some soft ticks have leg-associated glands that secrete clear coxal fluid. This secretion is thought to be separated plasma from the host upon which they fed. This potentially pathogen-laden fluid may further promote disease transmission in addition to salivary transmission. People may become infected when infected coxal fluid contacts breaks in the skin.

Medically important soft ticks

Genus *Argas*. These ticks are worldwide ectoparasites of birds, bats and reptiles that occasionally bite humans. About 60 species are known. They are distinctly flattened in appearance. Several species of *Argas* can transmit Q fever. The pigeon tick, *A. reflexus,* occurs in numerous European

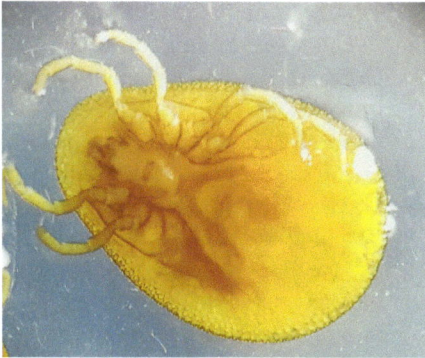

Fowl tick (*Argas persicus*). Source: Navid.drogba/ Wikipedia. CC BY-SA 4.0. Available from https://en. wikipedia.org/wiki/Argas_persicus#/media/ File:Argas_persicus.jpg

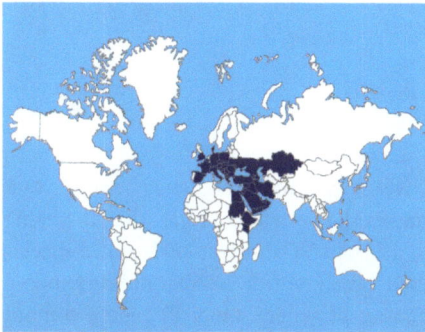

Distribution of *Argas* spp.

locations, where it infests pigeons and occasionally enters homes and bites people. Other notable pesky ticks in this genus are the African, *A. brumpti*, the largest known species, measuring up to 20 mm (0.8 inch) in length, *A. plonicus* and *A. vulgaris*.

Genus *Ornithodoros*. Most species occupy dry caves, burrows, stables, marine bird colonies, and beneath trees that shade livestock and wildlife. At least 10 species transmit spirochetes (*Borrelia* spp.) that cause tick-borne relapsing fever. *Ornithodorus moubata* is the most important vector of tick-borne relapsing fever, owing to its close human habitation association in

Ornithodorus moubata. Source: Jim Occi

Angola, south-west Africa, Botswana, Mozambique, Tanzania, Kenya, Zimbabwe and Madagascar. Other known relapsing fever vectors include: (1) *O. erraticus* (North Africa, Central Asia, Spain, Portugal, and southern Russia); (2) *O. tholozani* (Uzbekistan, Kashmir to Cyprus, Tripoli); (3) *O. verrucosus* (Caucasus); (4) the relapsing fever tick, *O. hermsi* (western United States, Canada); (5) *O. turicata* (western United States, Kansas to Mexico); (6) *O. parkeri* (western United States); (7) *O. rudis* (Panama, Colombia, Venezuela, Ecuador); (8) *O. talaje* (Mexico, Guatemala, Panama, and Colombia); and (9) *O. macrocanus* (south-western Europe, northwestern Africa). In addition, *O. chiropterphila* transmits Kyasanur Forest disease. Other annoying *Ornithodoros* species include *O. coriaceus* (Mexican tlalaja,

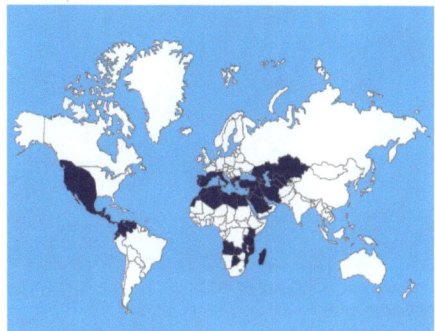

Distribution of *Ornithodoros* spp.

Californian pajaroello), which inhabits bedding areas used by deer and cattle, and it can inflict painful bites. The Spanish pigeon pest, *O. coniceps,* occasionally enters houses and bites people.

Genus *Otobius*. This genus contains two species but only one, the spinose ear-tick, *Otobius megnini*, is a nuisance to people. They are so named because they have a spiny cuticle. They do not transmit disease pathogens, but instead they can invade ear canals causing a painful, irritating infestation known as otoacariasis or parasitic otitis. They are unique among the soft ticks because the adults have a vestigial hypostome (mouthparts), so they do not feed. Only the larva and two nymphal stages are parasitic. Spinose ear ticks are one-host ticks, which naturally infest wild and domestic animals (wild: pronghorn antelope, mountain sheep, deer, elk; domestic: cattle, horses, mules, goats, sheep, dogs, rabbits). This species now has an essentially cosmopolitan distribution. Historically, they inhabited arid and semi-arid regions of western United States, Mexico and western Canada. However, livestock have spread them to western South America, Galapagos, Cuba, Hawaii,

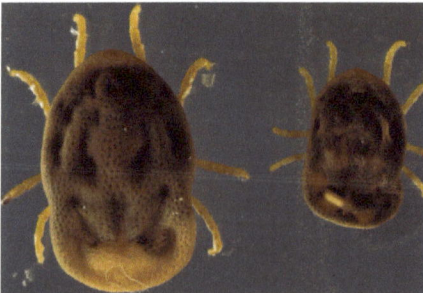

Spinose ear-tick (*Otobius megnini*). Source: University of Georgia-US Forest Service/Wikipedia. CC BY 3.0. Available from https://upload.wikimedia.org/ wikipedia/commons/8/8d/Otobius_megnini.jpg

south-eastern Africa, Madagascar and India. Following feeding, the engorged nymph leaves the host and hides under debris, rocks, and tree bark or in cracks and crevices, and subsequently moulting to the adult stage. Adult females live up to 2 years, laying 1500 eggs in several clutches. The hatching larvae crawl up anything vertical (vegetation, fence post, etc.) to ambush any passing host. The larva migrates rapidly along the body to the ear canal where it and the nymphs remain until they are through feeding. They feed 5 weeks to several months on their natural hosts. There can be one to two generations a year.

Hard ticks
(Family Ixodidae)

Compared with argasid ticks, ixodid ticks have a thicker, toughened cuticle that makes the body more rigid, thus the name 'hard ticks.' There are ~694 ixodid species distributed worldwide that are distributed among 14 genera. Only 10 genera contain medically important species, with the most important genera being *Ixodes*, *Amblyomma*, *Haemaphysalis*, *Hyalomma*, *Dermacentor* and *Rhipicephalus*. Similar to the argasids, a few ixodid species are confined to host burrows or nests (endophilic) and are seldom encountered by most people. However, most adult and immature ixodids are more widely scattered typically found attached on various hosts or searching for hosts on which to feed.

General biology

Adult and nymph hard ticks have eight walking legs (four pairs), and larvae or 'seed ticks' have six (three pairs). They lack antennae, and only have two body regions:

abdomen and cephalothorax (head and thorax combined), which are covered by a leathery cuticle. The front end (capitulum) bears the mouthparts containing sensory organs, cutting organs and an attachment/feeding structure (the hypostome) bearing numerous recurved teeth.

Ixodid ticks go through four life stages: egg, larva, nymph and adult. Both male and female ticks require a blood meal to progress (moult) to the next life stage and produce eggs. Depending on the species, the larvae, nymphs (three to five instars), adult stage, or all three require at least one blood meal. Therefore, all active stages, depending on species, can potentially bite. Most adult males feed briefly and sparingly, and some do not feed. Females must get enough blood to produce viable eggs. Most hard ticks are seasonal, attacking while environmental conditions are most suitable. Adult ticks are usually autumn (fall) and spring feeders, although some species of *Ixodes* may be active at temperatures just above freezing. Some ticks feed on only one host in their life, some feed on two hosts, but most hard ticks require three different hosts. Adult females typically lay their eggs in spring. Larvae emerge 2 to 4 weeks later and are generally most active in late spring/early summer or early autumn. Most nymphs are active during spring and summer. One complete life cycle may take 1 to 6 years.

A typical life cycle for a three-host hard tick is 2 to 3 years, but ranges from 6 months to 6 years, depending on environmental conditions, including temperature, relative humidity and photoperiod. However, tropical species can produce one to three generations a year. Hard ticks spend >90% of their life off their natural

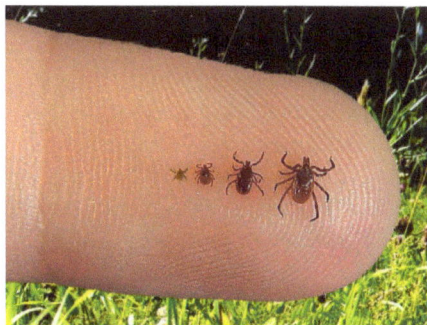

Some life stages of hard ticks left to right: larva, nymph, male, female. Source: California Department of Public Health/FLICKR/CC BY-ND 2.0. Available from https://www.flickr.com/photos/fairfaxcounty/7209178448/

hosts. The blood-fed females leave the third host to lay eggs, usually dropping to the ground. Egg development takes several days (3 weeks in tropics) before hatching into larvae. These 'seed ticks' grab a passing small animal (rodent, rabbit, etc.) or unsuspecting person and then feed on their blood for 2 to 7 days. Once engorged, they drop to the ground, and moult to the nymphal stage, which typically overwinter without feeding. Eventually, nymphs feed on another small animal or person, engorge and drop to the ground. They then moult to adults, which, again can overwinter without feeding. The adults then seek a large animal (e.g. coyote, deer, bear, cow, person) to find a mate and feed in order to produce viable, fertilised eggs. Male ticks consume relatively little blood, but they may remain on the host for a longer time and may mate with several females. The enormously swollen gravid females leave the host to seek a suitably warm and humid place to deposit a single batch of 400 to 20 000 eggs. Each stage (larva, nymph, adult) only takes one prolonged blood meal, but any stage may endure long starvation periods, sometimes years.

Ticks find potential hosts through a variety of means. Most ticks lack eyes, but they have various hair-like olfactory and gustatory sensory organs on the body, legs and mouthparts (Haller's organ). They use these structures to detect stimuli such as aromatic chemicals (carbon dioxide, octenol, ammonia and phenols), humidity, vibrations and body temperature. Hard ticks commonly climb to the very tips of grasses and other low vegetation to seek their next host where they initiate questing behaviour. They sway side-to-side sensing the environment while holding the first pair of legs out-stretched, like antennae, with the claws wide open. This is an ambush strategy, and it is the most common way hard ticks find their host. Ticks also may use a hunter strategy where they actively crawl towards the potential host. Some species use both strategies. Both nest and burrow dwelling soft and hard ticks use a third strategy: they simply hide until the host accidentally enters their habitat.

Ticks do not jump onto their host, but rather they quickly flex their claws to seize hair or clothing fibres. Under certain conditions when pheromones bring together

Questing larval ticks. Source: Jean and Fred/Flickr. CC BY 2.0. Available from https://www.flickr.com/photos/jean_hort/33291650492/

large aggregations of questing ticks, many ticks may attack unwitting people. Although many ticks feed on people, they are not always the preferred host. For example, *R. sanguineus* prefers dogs and other wild animals, but *I. ricinus*, *I. scapularis* and *A. hebraeum* readily feed on people. Before actually feeding, a tick may wander around on your body for several hours, doggedly climbing upward until they locate bare skin and a location where they can feed. Different species also prefer different body areas at which to attach and feed. For example: *D. variabilis* favours the head and neck (59%); *A. americanum* mainly favours the lower extremities, buttocks and groin (54%); *R. sanguineus* favours the head on children and the body on adults; and *I. scapularis* does not seem to have a preferred location. In general, most ticks attach to the head, neck and groin.

The other noteworthy aspect of tick biology is their incredible longevity. Even though all stages (particularly immatures) are subject to desiccation, ticks have behavioural and physiological adaptations to prevent excessive water loss. They can also endure long periods without food. Soft ticks, discussed earlier, are particularly

A questing tick (*Ixodes scapularis*). Source: US National Park Service

noteworthy; some species survive up to 16 years under starvation conditions.

Feeding

Ixodids are typically slow feeders taking 2 to 15 days per blood meal, and it may take up to 24 hours before a tick can properly attach and begin feeding. During feeding, the hypostome causes extensive tissue damage. They secrete a cementing material or plug around their hypostome in the bite wound that helps them obtain and maintain blood flow as they feed. Most hard tick bites are not painful and may go unnoticed. This is because they inject salivary juices containing anaesthetic, anti-inflammatory, antihaemostatic and immunosuppressive substances to deaden the wound, stimulate blood flow and create a blood pool. Host preference is also a key factor. Hard ticks can be: 'one-host' ticks such as the winter tick, *Dermacentor albipictus,* spending their larval, nymphal, and adult stages on a single host; 'two-host' ticks such as the tortoise tick, *Hyalomma aegptium*; or ' three-host' ticks such as *D. andersoni*. Most medically important hard ticks are 'three-host' ticks because multiple blood meals on multiple hosts increase the

Blood engorged deer tick (*Ixodes scapularis*). Source: Jerzy Gorecki/Pixabay. CC0. Available from https://pixabay.com/en/tick-lyme-disease-mites-bite-2371782/

chance of encountering and transmitting pathogens. Blood-fed ticks can be several times larger in size than their unfed counterparts.

Medically important hard ticks

Genus *Amblyomma*. About 102 species of *Amblyomma* inhabit primarily tropical and subtropical regions worldwide, although primarily in the Americas. The species occurring in the United States are the only temperate representatives. All *Amblyomma* appear to have a three-host life cycle, making them prime candidates for transmitting disease pathogens. Eight species of *Amblyomma* are known to carry 14 different viruses. The African tropical bont tick *A. variegatum* may carry five viruses including Dugbe virus, Bhanja virus, Thogoto virus, lymphocytic choriomeningitis virus, and Crimean-Congo haemorrhagic fever virus. The South African bont tick, *A. hebraeum*, can carry and transmit African tick-bite fever. Other species of *Amblyomma* transmit spotted fever group rickettsia (tick-borne typhus). For example, the Cayenne tick, *A. cajennense*, transmits tick-borne typhus in Mexico, Panama, Colombia, and Brazil. The lone star tick, *A.*

A lone star tick (*Amblyomma americanum*) attached for feeding. Source: NIAID/Flickr. CC BY 2.0. Available from https://www.flickr.com/photos/niaid/37280298750/

African tropical bont tick (*Amblyomma variegatum*). Alan R. Walker/Wikipedia. CC BY-SA 3.0. Available from https://en.wikipedia.org/wiki/Amblyomma_ variegatum#/media/File:Amblyomma-variegatum-male.jpg

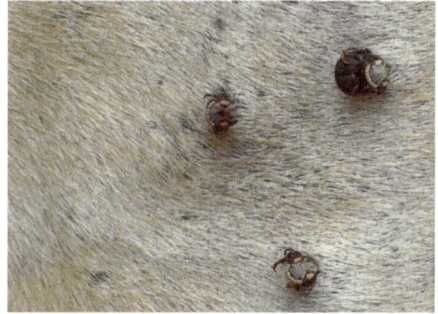

South African bont tick (*Amblyomma hebraeum*). Source: Bernard DuPont/Flickr. CC BY-SA 2.0). Available from https://www.flickr.com/photos/ berniedup/16486455419/

americanum, transmits tularemia and may be a secondary vector of Rocky Mountain spotted fever. Other *Amblyomma* ticks carry Botonneuse fever, another rickettsial

Lone star tick (*Amblyomma americanum*). Source: James Gathany, US Centers for Disease Control and Prevention

disease. The bite of the lone star tick can infect its victim with a salivary compound that can cause allergies triggered by eating mammalian meat (Alpha-gal allergy), which is being increasingly reported in the United States, particularly the south-eastern region. The symptoms of this allergy are complex typical of most food allergies, and may include anaphylaxis. Finally, many rural and forest dwelling *Amblyomma* in Asia and the Americas are aggressive biters and can be a major nuisance.

Genus *Dermacentor*. The 31 known species of *Dermacentor* inhabit temperate zones worldwide, except Australia where they are absent. The Rocky Mountain wood tick, *D. andersoni*, alone can carry and transmit the pathogenic agents of five diseases, including Rocky Mountain spotted fever, tularemia, Powassan encephalitis virus, Colorado tick fever and lymphocytic choriomeningitis. In addition, *D. andersoni* and the American dog tick, *D. variabilis*, are known to cause tick paralysis. The Pacific coast ticks, *D. occidentalis* and *D. variabilis*, can transmit tularemia. In addition, *D. occidentalis* transmits Colorado tick fever and Rocky Mountain

Rocky Mountain wood tick (*Dermacentor andersoni*). Source: Andrey Zharkikh/Flickr. https://www.flickr.com/photos/zharkikh/6806832735/

Rabbit tick (*Haemaphysalis leporispalustris*). Source: George Schultz, US Department of Defense

American dog tick (*Dermacentor variabilis*). Source: David E. Bowles

spotted fever. The rabbit tick, *D. parumapertus,* also transmits Rocky Mountain spotted fever. In western Siberia, both *D. pictus* and *D. marginatus* transmit and serve as reservoirs of Omsk haemorrhagic fever virus. In Europe, *D. marginatus* and *D. reticulatus* are the vectors of tick-borne encephalitis, while in India *D. auratus* transmits Kyasanur Forest disease. In Asia, *D. taiwanensis* transmits oriental spotted fever. Other species of *Dermacentor* transmit North Asian tick typhus, Q fever and boutonneuse fever.

Genus *Haemaphysalis*. The 150 species of *Haemaphysalis* primarily inhabit Old World temperate and tropical zones but five species occur in North and South America. There are several important disease vectors in this genus. The European winter tick, *H. inermis,* transmits tick-borne encephalitis while *H. punctata* is a vector of Bhanja virus in Italy. In India, *H. spinigera* and *H. bispinosa* are vectors of Kyasanur Forest disease, *H. intermedia* transmit Bhanja virus, and *H. intermedia* and *H. wellingtoni* transmit Ganjam virus. Three species of *Haemaphysalis* serve as vectors of Siberian tick typhus (North Asia tick typhus), and *H. longicornis* is a vector of Powassan encephalitis virus in Russia's southern Maritime Province. The yellow dog tick, *H. leachi,* (Africa, Asia) transmits boutonneuse fever (African tick typhus, Kenya tick typhus). The New World rabbit tick, *H. leporispalustris* (Alaska and Canada to Argentina), bites people only infrequently and is not considered a human disease vector, but it plays an important role in maintaining the Rocky Mountain spotted fever rickettsiae among wild animals, as well as tularemia.

Genus *Hyalomma*. *Hyalomma* inhabit unforested steppes, savannas and deserts from Eurasia south through the Middle East, Arabia and Africa. There are 21

Hyalomma marginatum rufipes (female left, male right). Source: Daktaridudu/Wikipedia. CC BY-SA 4.0. Available from https://en.wikipedia.org/wiki/Ticks_of_domestic_animals#/media/File:Hyalomma-rufipes-female-male.jpg

known species some of which travel on, and are widely distributed by, migratory birds. Others are associated with nomadic camel caravans, and migratory wildlife. These distributional behaviours facilitate their wide dissemination of a variety of tick-borne diseases. Twelve species of *Hyalomma* are known to carry over 17 different arboviruses. In southern and eastern Europe, *H. marginatum marginatum* and *H. anatolicum anatolicum* attack people more aggressively than other species and they are competent vectors of Crimean-Congo haemorrhagic fever. The potentially fatal Thogot virus is carried by *H. a. anatolicum* in Egypt, while *H. impeltatum* and *H. marginatum isaaci* transmit Wanowrie virus in India. The Nigerian species *H.*

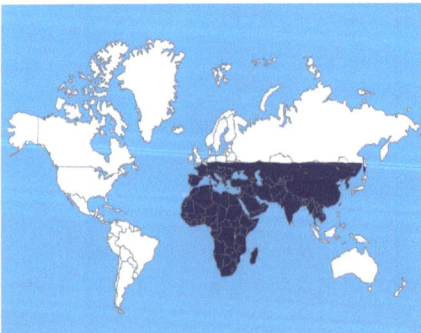

Distribution of *Hyalomma* spp.

marginatum rufipes transmits Dugbe virus and boutonneuse fever (South African tick typhus). Bhanja virus is transmitted by *H. truncatum* in Nigeria and Cameroon. Finally, *H. asiaticum* is a vector of Siberian tick typhus in Kirgizia.

Genus *Ixodes*. This is the largest ixodid genus including ~243 species occurring in humid arctic, temperate, subtropical and tropical habitats worldwide. Most species in this genus closely resemble one another in having dark brown to black legs and scutellum and reddish-coloured bodies. Besides being annoying biters and causing tick paralysis, they are known to transmit eight bacterial and eight viral diseases. Some species are known to be vectors of multiple pathogens. For example, the European castor bean tick, *Ixodes ricinus*, is particularly noteworthy. It is found throughout Europe, northern Africa and northern Asia, and it is a competent vector of Lyme disease, human granulocytic ehrlichiosis, human babesiosis, tick-borne encephalitis, louping ill and Tribec virus. The black-legged tick, *I. scapularis*, is also a formidable disease vector being a vector of Lyme disease, human babesiosis and human granulocytic ehrlichiosis throughout the east and Midwestern United States. Similarly, the western black-legged tick, *I. pacificus* transmits Lyme disease in the western United States. Another noteworthy species is the taiga tick, *I. persulcatus*, which transmits Lyme disease, Omsk haemorrhagic fever and Kemerovo virus in northern Asia, central and eastern Europe. A related species, *I. pavlovskyi*, transmits tick-borne encephalitis (Russian spring-summer encephalitis) in those same areas. In Japan, South-East Asia, India, China and Taiwan, *I. granulatus* transmits human

Black-legged tick (*Ixodes scapularis*). Source: US Department of Agriculture/Flickr. CC BY 2.0. Available from https://www.flickr.com/photos/usdagov/8456915056/

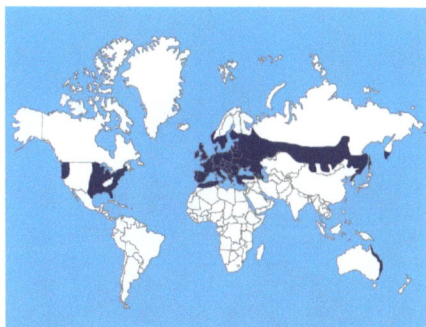

Distribution of primary *Ixodes* spp. that feed on humans.

babesiosis, Langat virus and tick-borne typhus. Kyasanur Forest disease is transmitted by *I. petauristae* and *I. ceylonensis* in India. Many other species of *Ixodes* are known or suspected vectors of disease agents. Other species serve as nuisances or are painful biters such as the worlds' largest tick, *I. acutitarsus,* distributed from the Himalayas to southern Japan. Still others including *I. redikorzevi* (Israel), *I. holocyclus* (Australia) and the Karoo paralysis tick, *I. rubicundus,* (South Africa) may cause tick paralysis.

The Australian paralysis tick, *I. holocyclus*, produces a salivary compound that can cause a severe allergic reaction in

Australian paralysis tick (*Ixodes holocyclus*), female left, male right. Source: Stephen L. Doggett

bitten humans that is triggered when they subsequently eat mammalian meat. This Mammalian Meat Allergy (MMA), also called Alpha-gal allergy, is nearly identical to the syndrome caused by saliva (bites) of the Lone Star tick in the United States (see earlier). As stated previously, the symptoms of this allergy are complex, similar to other known food allergies, and may include anaphylaxis. Australia has had numerous reports of this effect recently, nearly all associated with *I. holocyclus* bites. This syndrome has also recently been reported, but only rarely, from other countries, but the most likely associated causal ticks have not yet been clearly determined.

Genus *Nosomma*. This genus contains one human-biting tick, *Nosomma monstrosum,* which normally infests buffalo, domestic cattle and other wild animals throughout India, Bangladesh, South-East Asia, Nepal and Sri Lanka. This species is thought to be a potential vector of Kyasanur Forest disease.

Genus *Rhipicephalus*. About 50 *Rhipicephalus* species are native to Africa and 20 additional species are native to southern Europe, Asia, Indonesia and Ethiopian. The brown dog tick, *R. sanguineus*, is

Brown dog tick (*Rhipicephalus sanguineus*). Source: US Centers for Disease Control and Prevention, J. Gathany & W. Nicholson

Rhipicephalus pulchellus (male). Source: Daktaridudu/Wikipedia. CC BY-SA 4.0. https://en.wikipedia.org/wiki/Rhipicephalus_pulchellus#/media/File:Rhipicephalus-pulchellus-male.jpg

distributed worldwide largely due to transport of domestic dogs. Thirteen species of *Rhipicephalus* transmit six major viral and four bacterial diseases. Most notably, *R. sanguineus* transmits boutonneuse fever, lymphocytic choriomeningitis, tularemia (rabbit fever), Crimean-Congo haemorrhagic fever, and Siberian tick typhus (North Asian tick typhus) in the Old World, and it is a vector of Rocky Mountain spotted fever (tick-borne typhus) in the south-western United States. Others species also transmit boutonneuse fever (*R. pumilio),* Kyasanur Forest disease (*R. turanicus),* Thogoto virus (*R. bursa*) and Nairobi sheep disease (*R. appendiculatus).* Some species such as immature *R. pulchellus* (East Africa) are irritating pests and their bites may produce skin ulcers in some people.

Tick-borne disease prevention and treatment

Situational awareness and effective personal protection measures are key to avoiding tick problems (see 'Personal protection measures'). Because soft ticks and many hard ticks favour animal dwellings (e.g. burrows, caves, nests) that serve as habitat for their natural hosts, try to avoid such areas if possible. Remember soft ticks

Rhipicephalus pulchellus (female). Source Daktaridudu/Wikipedia. CC BY-SA 4.0. https://en.wikipedia.org/wiki/Rhipicephalus_pulchellus#/media/File:Rhipicephalus-pulchellus-female.jpg

are nocturnal, fast-feeding ticks that do not attach for long periods to feed. If camping, select campsites well away from rodent burrows, inspect cabins and other temporary or infrequently used dwellings for signs of rodents and birds. If avoiding such places is not possible, emphasis should be placed on personal protection measures to minimise risk of attack. Most hard ticks are 'edge dwellers' or 'edge species', so when outdoors avoid transitional vegetation along where woodlands transition to grasslands or brushy areas. Where practical, enclosed sleeping quarters are best (i.e. closed doors and windows, window/tent screens; sealed tent floors and zippered openings.).

Repellents can also substantially increase protection. That said, ticks easily crawl under lose fitting clothing so applying repellent only to exposed skin is generally an inadequate defence to protect against tick bites. Proper wear of protective clothing is crucial (see 'Personal protection measures'). Because ticks grab onto the host from a resting point on vegetation and then climb the body, a highly practical means of preventing them from attaching is to wear long pants tucked into footwear or socks. This simple arrangement prevents ticks from crawling under the pant legs and onto the skin. Wearing light-coloured clothing facilitates finding the ticks and removing them. In addition, wearing permethrin-treated clothing is an important tick prevention tool.

After returning from outdoors, carefully inspect your body for crawling or attached ticks and remove them. In some cases, the larvae and nymphs are very small (pinhead size) and can be difficult to detect. In addition, because ticks can attach anywhere and many bites are generally painless, they may go unnoticed. A hot bath or shower helps remove crawling ticks, and clothing should be changed and washed. Children and pets should similarly be inspected for ticks. When an attached tick is found, it should be removed immediately. Follow the procedures and precautions described in the next section for proper tick removal. The longer a tick remains attached, the more blood engorged it becomes, the more difficult it is to remove, and the more likely it is to transmit disease. Follow the guidance presented, collect any ticks you find on you and save them in a sealed container or zip-lock plastic bag. Alternatively, a high-quality photograph of the tick may serve this purpose. The tick specimen or photo should be given to your medical provider should you later become sick. Always inform attending medical personnel that you visited a high-risk area, or had contact with ticks, because this will aid them in their differential diagnosis of your condition. It is possible that part of the tick's mouthpart may remain in the skin following removal.

Proper attached tick removal procedures

How to remove attached ticks (per US Centers for Disease Control and Prevention guidance):

- Use medium-tipped forceps (sold as various commercially available 'tick removal devises'), place tips around mouthparts where they enter the skin.
- Slowly and gently pull the tick away from the skin or slide forceps along the skin (follow device directions). Do not jerk, crush, squeeze or puncture the tick.

- Directly place tick in sealable container.
- Disinfect the bite site using standard procedures.
- If possible, save the tick alive for identification and disease testing. Place it in labelled (date, your name, geographic location, etc.), sealed bag or vial with a lightly moistened paper towel or even a few grass blades then store at refrigerated temperature if possible.
- If forceps are unavailable, use index finger and thumb with rubber (latex) gloves, plastic or even paper towel to prevent finger contamination. Tick faeces can contain pathogens so bare-fingers can transfer these pathogens to cuts, abrasions, and nasal or eye mucous membranes.
- **Do not** apply petroleum jelly (Vaseline®), fingernail polish or similar chemicals over ticks, burn them or use various commercial gadgets to detach them. Such methods usually do not work, and may crush, squeeze or cause the tick to regurgitate and thus increase disease transmission and/or infection.

Preventing tick attachment and feeding, and rapid removal of attached ticks are crucial because of the many tick-borne pathogens carried by these animals

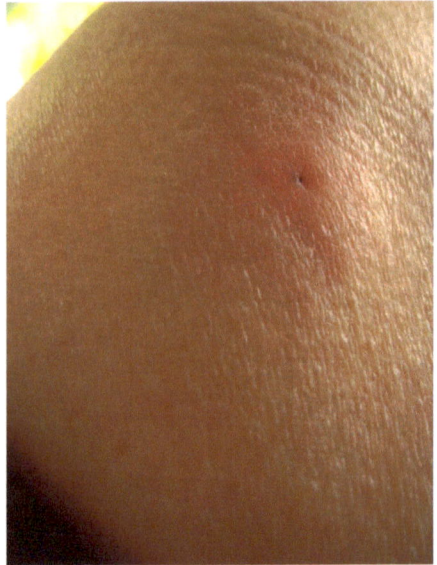

Tick mouthparts left in skin following removal. Source: KitAy/Flickr. CC BY 2.0. Available from https://www.flickr.com/photos/kitpfish/1910223778/

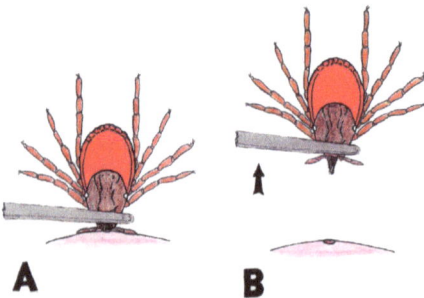

Proper attached tick removal procedure. Source: Harold Harlan

that may cause serious diseases. Some vaccines are available and may be appropriate for some travellers or other people who routinely contact high-risk tick habitats (see the 'Education' section resources and Appendix 7). Additionally, some antibiotics may serve as prophylactics for bacterial pathogens, but, similar to vaccines, they are only prescribed when your travel/adventure destination and objectives are likely to expose you to high-risk tick-borne diseases. Vaccines and prophylactic medication take time to become effective so see your health-care provider or travel medicine clinic at least 6 weeks before going to high-risk areas. Regardless, no vaccine or prophylactic drug is 100% effective, so using personal protection measures is essential to protect health. After returning from travel to areas with high risk of tick-borne disease transmission, you should be alert for any onset of illness and should seek prompt medical attention as needed. Bacterial tick-borne diseases are usually effectively treated using broad-spectrum antibiotics, but viral diseases often cannot be specifically treated with medicines and require supportive care. Because viral diseases are difficult to treat, it is imperative that medical assistance be provided at the earliest opportunity following onset of illness.

Centipedes and millipedes
(Classes Chilopoda and Diplopoda)

Centipedes
(Class Chilopoda)

The 3000 known centipede species inhabit temperate, subtropical and tropical terrestrial habitats worldwide. Most are small, but some tropical species reach nearly 60 cm (2 feet) in length. They are multi-legged (one pair per segment), elongate, flattened animals, and they can move rapidly. The first pair of legs, called toxicognaths, have venom glands and are modified piercing claws for capturing prey. Most centipedes are completely harmless, although some larger bodied species can inflict painful bites. The last pair of legs can also inflict a mild, non-venomous pinch. The true mouthparts are too small to bite through skin, and some centipedes secrete an offensive yet harmless substance for repelling predators. Normally secretive, centipedes are very agile and can move rapidly. They typically seek shelter under rocks, rotten logs and loose tree bark where they prey on smaller organisms.

The centipede genus *Scolopendra* has roughly 80 subtropical and tropical species distributed worldwide. These are the largest and most venomous centipedes, and their bites can cause severe pain lasting up to several hours. The bites of some species can kill a mouse within 30 seconds. Reactions to their bites include localised tissue swelling, redness, swollen painful lymph nodes, headache, heart palpitations, respiratory distress, nausea and/or vomiting, anxiety, and rarely death. Tissue necrosis around the bite wound is common,

Giant centipede (*Scolopendra heros*). Source: John/ Flickr CC by 2.0. Available from https://www.flickr. com/photos/8373783@N07/6884438698

but secondary infection is almost unknown. Symptoms and signs from the bite seldom persist more than 48 hours. Numerous tropical and subtropical centipedes in the genus *Otostigmus* exude defensive chemical compounds from glands along the body. These compounds are usually not toxic to humans, but they may cause formation of skin vesicles upon contact. Situational awareness and avoidance are crucial for avoiding unwanted contacts from centipedes. Should a bite occur, the wound should be thoroughly

Giant centipede (*Scolopendra heros*), dark form. Source: David E. Bowles

Head of *Scolopendra* sp. showing fang-like toxicognaths. Source: Harold J. Harlan

Millipede in a defensive coil. Source: Jane Schlossberg/Flickr. CC BY-SA 2.0. Available from https://www.flickr.com/photos/jschlossberg/5735047097/

cleaned followed by application of topical corticosteroids, use of systemic antihistamines and cold compresses. Severe reactions to the bites from centipedes may require immediate medical attention, and they should be evaluated on a case-by-case basis.

Millipedes
(Class Diplopoda)

Millipedes superficially resemble centipedes, but they have two pairs of legs per body segment, and cylindrical bodies. Their body size ranges from near microscopic up to 30 cm (1 foot) long in some tropical species. Millipedes inhabit various terrestrial environments, but they prefer damp habitats, including underground shelters, leaf litter, soil and rotting wood. Unlike centipedes, millipedes are sluggish and generally slow moving. Most millipedes are completely harmless, secretive and do not bite. However, when alarmed, some species roll into a coil and excrete noxious chemicals from pores located along their sides. Potency of these chemicals varies among species, but the odour is usually offensive. Moreover, some tropical millipedes can eject chemical excretions up to a metre (1 yard) away,

which can burn your eyes and skin on contact. Skin reactions may include yellow or brown staining, intense burning, itching and occasionally blistering. Severe cases result in markedly reddened skin and necrosis. In most cases, symptoms usually resolve within 24 hours, but contamination of the eyes may cause severe conjunctiva reactions with corneal ulceration and recovery may take several days in those cases. Particularly injurious species include *Rhinocricus lethifer* and *R. latespargor* (Haiti), *Polyceroconas* spp. (Papua New Guinea), *Spirostreptus* spp. and *Julus* spp. (Indonesia), *Spirobolus* spp. (Tanzania), *Orthoporus* spp. (Mexico) and *Tylobolus* spp. (California).

Like centipedes, situational awareness, and avoidance are the keys to preventing

A tropical millipede. Source: David E. Bowles

Distribution of injurious millipedes.

exposure to millipede toxins. Should exposure occur, applying copious amounts of rubbing alcohol or soap and water can be used to wash the toxin from the skin. Use of topical corticosteroids, systemic antihistamines and antibiotics are useful for treating skin reactions. Eye injuries require immediate irrigation followed by immediate professional medical care. Common treatments for eye burns include corticosteroid eye drops or ointments.

Crustacea

The Class Crustacea includes crabs, lobsters, shrimp and crayfish, as well as many other smaller taxa. Numerous crustaceans, mostly in the Order Decapoda, grow large bodies with claws (pinchers) and body spines that can inflict injury. Large crabs, lobsters, freshwater prawn and crayfish can inflict painful pinches. Large stone crabs can cause serious cuts and amputated fingers when they are mishandled. Several species of stone crabs in coastal areas of southern and south-eastern Asia have been reported to feed on various jellyfish and/or dinoflagellates, thus ingesting saxitoxins or tetrodotoxins in sufficient amounts to cause serious poisoning of persons who later cooked and ate those crabs (especially members of the crab Family Xanthidae). Among the Crustacea,

Mantid shrimp (*Odontodactylus scyllarus*). Source: Roy Caldwell, US National Science Foundation

mantid shrimps are perhaps best known for causing injuries to people, and they are sometimes referred to as 'thumb splitters.' Some mantid shrimps can grow over 30 cm (~1 foot) long, and these sometimes aggressive, predatory crustaceans, may use their formidable claws to inflict cuts on the hands of those who try to handle them. The claws are so powerful that they have been known to break aquarium glass with their thrusts. Handling a mantid shrimp without protective gloves may result in serious injury.

Lobster (*Homarus gammarus*). Source: Richie Rocket/Flickr. CC BY-ND 2.0. Available from https://www.flickr.com/photos/richierocket/36130917314/

Stone crab (*Menippe mercenaria*). Source: Andera Westmoreland/Flickr. CC BY-SA 2.0). Available from https://www.flickr.com/photos/andrea_pauline/4571204589/

Lice
(Order Phthiraptera)

Two sucking louse families, Pediclulidae and Pthiridae, are exclusively human parasites and occur wherever people live worldwide. Lice are wingless, variously coloured (usually grey, brown or black), with prominent tarsal claws, and ~2.5–3.5 mm (0.09 to 0.13 inch) long. Their bites are primarily a nuisance, and they cause reddened skin patches, lesions, wheals and severe itching. Scratching may result in secondary infections. Head and pubic lice attach or cement their eggs (nits) to hairs, while body lice primarily attach their eggs to clothing or bedding. Nits hatch in only a few days, allowing for rapid population growth. Under crowded and/or unsanitary living conditions (natural disasters, refugee camps, etc.), louse infestations spread rapidly from person to person. Successful control of louse infestations, involves killing the adults and nymphs, as well as removing their nits.

Head lice
(Pediculus humanus capitis)
The head louse exclusively infests human hair. Anyone is susceptible, but particularly children, refugees, prisoners of war, concentration camp detainees, vagrants and others in poor socio-economic or stressful situations. Transmission is via shared clothing (hats/caps, scarfs, hoodies, burkas, etc.), combs/brushes, pillows/bedding, direct head to head contact, and so on. Permethrin-based shampoo or louse (nit or 'fine tooth') combs are the usual hair treatments. Dry-cleaning, very hot washing with detergent, hot air drying (1

Head louse (*Pediculus humanus capitis*). Source: David E. Bowles

hour), and various chemicals (pesticides) kill lice and nits. Head lice are secondary vectors of epidemic typhus, but the threat from them is largely insignificant.

Body lice
(Pediculus humanus humanus)
The body louse lives and lays eggs on clothing seams or bedding material, and normally only contacts human skin to feed. Body lice are the primary epidemic typhus, louse-borne relapsing fever, and trench fever vectors. They often infest refugees, prisoners of war, concentration camp detainees, vagrants and others with poor hygiene. Under such conditions, typhus outbreaks can be explosive and difficult to control. Control of louse populations can be achieved by washing clothing and bedding materials with hot water and detergent washing and hot air drying (1 hour). Heavy infestations, particularly

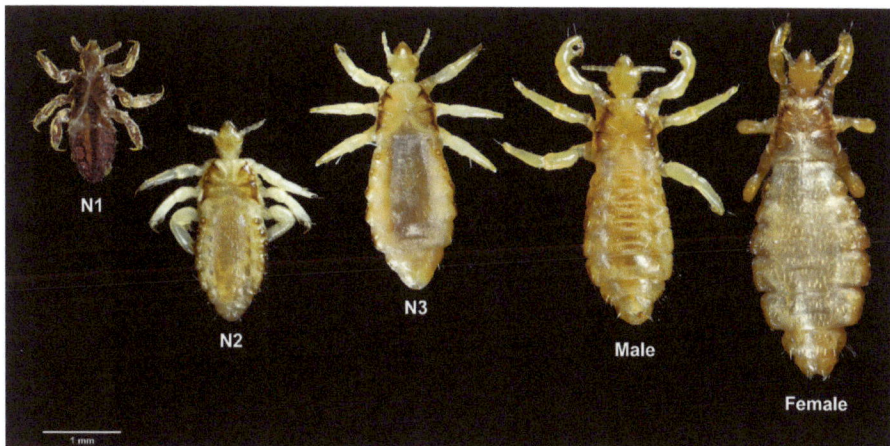

Head louse (*Pediculus humanus capitis*) life stages. Source: James Gathany, US Centers for Disease Control and Prevention

among refugees, detainees and prisoners of war, may require insecticide applications to the body, as well as clothing and bedding.

Pubic lice
(*Pthirus pubis*)

The pubic louse or crab louse primarily infests pubic hair, but beards, eyelashes and eyebrows may also be involved. They are transferred person-to-person primarily through sexual contact. Their presence on the body may cause psychological distress in some people, but they do not transmit any known human diseases. Permethrin-based topical ointments or shampoo is the usual effective treatment. Pubic lice do not live long when they are not on humans. Following treatment of infested individuals, any straggler lice can be effectively controlled through sanitation of clothing and bedding including dry-cleaning, in hot water and detergent washing, and hot air drying (1 hour).

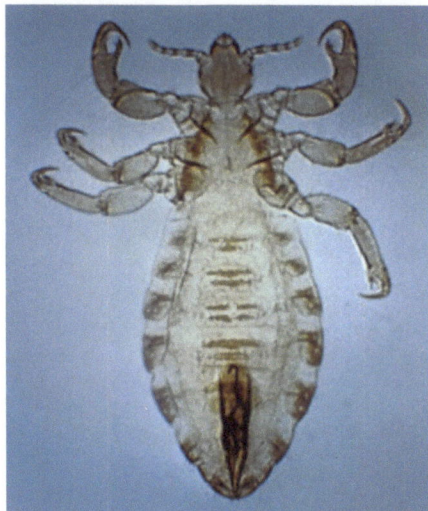

Body louse (*Pediculus humanus humanus*). Source: US Centers for Disease Control and Prevention

Pubic louse (*Pthirus pubis*). Source: Stephen L. Doggett

Cockroaches
(Order Blattodea)

German cockroach (*Blattella germanica*), female. Source: Harold J. Harlan

German cockroach (*Blattella germanica*), male. Source: Harold J. Harlan

Several thousand species of cockroaches occur worldwide, but only a few pest species are closely associated with human habitations. All of the pest cockroach species have cosmopolitan distributions. These include the German cockroach (*Blattella germanica*), brown-banded cockroach (*Supella longipalpa*), smoky brown cockroach (*Periplaneta fuliginosa*), Asian cockroach (*Blatella asahinae*) and oriental cockroach (*Blatta orientalis*). The American cockroach (*Periplaneta americana*) is an incidental entrant into homes in the southern United States and Mexico, but it is not a significant pest. Other species also occasionally enter houses, but they

Oriental cockroach (*Blatta orientalis*). Source: Harold J. Harlan

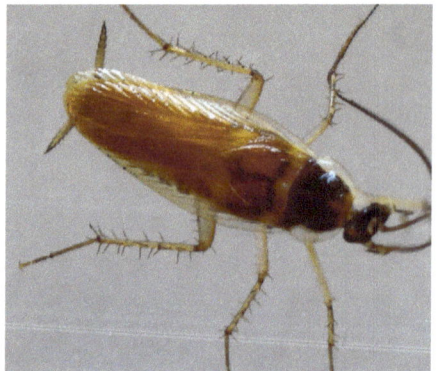

Brown-banded cockroach (*Supella longipalpa*). Source: Harold J. Harlan

American cockroach (*Periplaneta americana*). Source: Alex Wild/Flickr. CC0. Available from https://www.flickr.com/photos/131104726@N02/25859665686/

also are not considered significant pests. Strictly speaking, cockroaches are not medical pests and normally are not a major threat to human health. However, their very presence may cause psychological distress and lowered morale in some people. Similarly, although uncommon, high cockroach populations and frequent exposure to cockroach body parts and faeces sometimes causes allergic reactions in some people. Microbes carried on the bodies of cockroaches can also contaminate food preparation surfaces and unprotected food.

True bugs
(Order Hemiptera)

Many true bugs have piercing-sucking mouthparts and can inflict painful bites. Some can transmit potentially deadly parasitic diseases.

Kissing bugs
(Family Reduviidae, Subfamily Triatominae)

There are ~138 species in this subfamily that are mainly distributed from the southern United States southward throughout South America. All medically important triatomids feed on mammalian blood for augmenting reproduction. Among them, a few species can transmit the parasites (*Trypanosoma cruzi*) that cause Chagas disease, or American trypanosomiasis. The bites they inflict while feeding are relatively painless, which allows them to feed on their hosts without disturbing them. However, some species can also deliver a painful bite for defensive purposes. Most feeding bites by kissing bugs occur at night when the victim is sleeping and the bites generally occur on the face. Kissing bugs are nocturnal and they retreat to wall or ceiling cracks, under floors or wall hangings, and into thatched roofs to hide during the day, emerging nightly to feed while you sleep. A classic reaction to the bite of a kissing bug is formation of a chagoma, which is an inflammatory nodule. The chagoma typically is a hard, swollen, violet-hued, lesion with a purplish-coloured puncture mark. In addition to a chagoma, some individuals may develop localised urticarial reactions to kissing bug saliva, low blood pressure, itching, vomiting, headache and abdominal cramping. Women may experience uterine bleeding.

The transmission of the Chagas disease parasites does not occur through the bite of the kissing bug. Rather, while the bug feeds it leaves parasite-contaminated faeces on the skin, which subsequently enters the bite wound or cuts often due to the bitten host scratching the bite site. The kissing bug is typically already gone when such infection occurs. An additional route of infection is from bugs resting in their hiding places that drop contaminated faeces on the victim. If the insect is hiding above a person while they are sleeping, the

Kissing bugs (left to right, *Triatoma protracta*, *T. gerstaeckeri*, *T. sanguisuga*). Source: Curtis-Robles *et al.*/Wikipedia, CC BY 4.0 International. Available from https://commons.wikimedia.org/wiki/File:Three_species_of_kissing_bugs.PNG

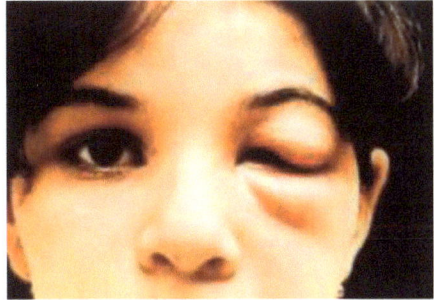

Romana's sign. Source: World Health Organization

Kissing bug (*Panstrongylus geniculatus*), Panama. Source: David E. Bowles and Mark Pomerinke, United States Air Force

faeces may contact the eyes, mucous membranes or open mouth, which can lead to infection. When the parasites enter through the conjunctiva of the eye, it may result in a unilateral swelling of the lymphoid tissue known commonly as Romana's sign. Consuming food or drinks contaminated by infective bug faeces, plus blood transfusions and transplantation of infected organs are also effective routes of human infection.

Treatment of kissing bug bites includes use of cool compresses and mild analgesics to relieve the itching. Occasionally, patients who are hypersensitive to kissing bug bites may develop severe allergic reactions, which are treated like any other severe allergic reaction. For individuals who demonstrate sensitivity to bites, immunotherapy can be beneficial in the long term.

Chagas disease is a serious disease. Treatment is complicated and must be conducted by medical professionals, so we do

Kissing bug (*Rhodnius prolixus*). Source: Erwin Huebner/Wikipedia. Available from https://upload.wikimedia.org/wikipedia/commons/b/bd/Rhodnius_prolixus70-300.jpg

Distribution of kissing bugs (Triatominae).

not cover it here. It is highly recommended that travellers consult the CDC, PAHO or WHO websites to determine if Chagas disease is a risk where they are going. Use of a permethrin-treated bed net is essential in high-risk areas. Finally, travellers to areas where Chagas disease is endemic should seek medical attention should any symptoms develop, particularly a chagoma or Romana's sign. Any chronic symptoms or unexplained health issues should be brought to the attention of a physician with an indication of travel history to a Chagas disease endemic area. Doing so will aid greatly in the differential diagnosis.

Assassin bugs
(Family Reduviidae, Subfamily Harpactorinae)

Unlike kissing bugs, assassin bugs are predators and not blood feeders, and they do not transmit any known human diseases. This large group of ~7000 species are distributed worldwide. Their defensive bites, which occur during accidental contact or handling, are extremely painful. Initially, the severe pain may last a few hours, and it is followed by residual pain and numbness lasting up to several days.

Masked hunter (*Reduvius personatus*). Source: Donald Hoborn/Flickr. CC BY-SA 2.0. Available from https://www.flickr.com/photos/dhobern/19287605490/

The afflicted area often becomes reddened and hot to the touch. It may turn white with a hardened core that can subsequently slough off leaving a small, puncture site hole. Healing time varies from 2 weeks to 6 months in some cases, especially in hypersensitive people.

Attacks by assassin bugs are relatively common on all continents. The most commonly recognised human-biting assassin bug in North America is the wheel bug, *Arilus cristatus*, but several closely related species of wheel bugs occur throughout Central and South America.

Wheel bug (*Arilus cristatus*). Source: John Flannery/Flickr. CC BY-SA 2.0. Available from https://www.flickr.com/photos/drphotomoto/3673429413

Assassin bug (*Pristhesancus plagipennis*). Source: James Niland/Flickr. CC BY 2.0. Available from https://www.flickr.com/photos/bareego/5289803034/

The masked hunter, *Reduvius personatus*, of North America, sometimes bites humans. Due to the red and black colouration of this species and a tendency for it to bite on the face, it is sometimes incorrectly identified as a kissing bug. *Pristhesancus plagipennis* has been implicated in several painful bites in Australia.

Situational awareness, avoidance and personal protection measures are key to avoiding assassin bug bites. Use of ice packs, topical corticosteroids and systemic antihistamines are effective for relieving pain/itching/swelling. Apply antibiotic ointment to minimise possible secondary infection.

Bat bug (*Cimex adjunctus*). Source: David E. Bowles

Bed bugs
(Family Cimicidae)

Although bed bugs (*Cimex lectularius* and *C. hemipterus*) do not transmit any human diseases, all the life stages except the eggs require blood meals to survive. When they feed, they use piercing-sucking mouthparts to inject saliva, which stimulates blood flow and deadens the bite. The initial bite is usually painless, but the saliva can cause intensely itchy, inflamed wheals. Some individuals, repeatedly exposed to the saliva, develop allergic reactions such as dermatitis, localised inflammation, and prominent wheals. Rarely, others become hypersensitive to the saliva and may develop asthma, urticaria, blisters, arthralgia and anaphylaxis, but such responses usually cease once the exposure ends. Persons with no, or limited, prior bites by bed bugs may have a delayed itchy red wheal at bitten sites, but this may not show up for 14 or more days after the actual bites occurred. Normally, topical corticosteroids and systemic antihistamines effec-tively manage the itching. Scratching the bites may lead to secondary infections.

Bed bugs are small, ≤0.24 inch (≤6 mm), oval, dorsoventrally flattened and reddish-brown in colour. They

Bed bug (*Cimex lectularius*). Source: Harold J. Harlan

Bed bug (*Cimex lectularius*) feeding. Source: P. Naskrecki, US Centers for Disease Control and Prevention

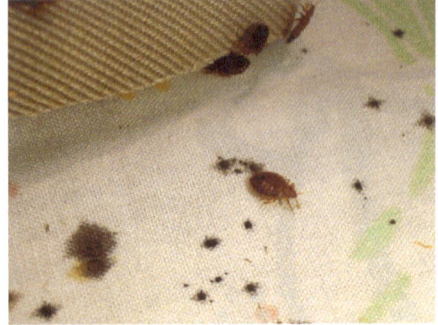

Bed bugs and faecal spots on bed linen. Source: Harold J. Harlan

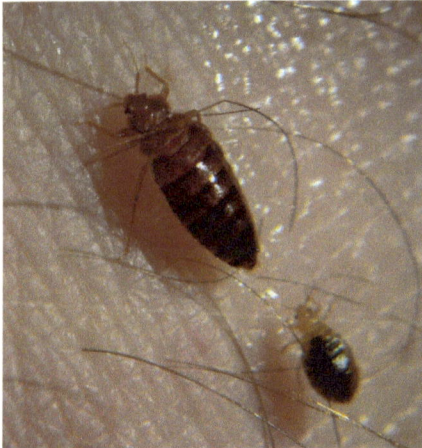

Bed bugs (*Cimex lectularius*), adult and nymph. Source: Harold J. Harlan

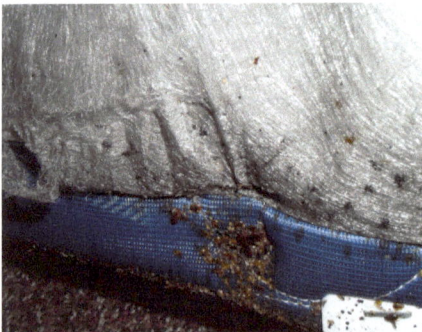

Beg bugs on a mattress. Source: Harold J. Harlan

usually hide during the day and emerge at night to feed on sleeping victims. Their primary harbourage is mattresses and other bedding but, during heavy infestations, they also hide behind picture frames, headboards, under carpet, behind wallpaper, and in wall cracks and crevices. Some species of Cimicidae also infest bats and/or some bird nests. Bed bugs are particularly problematic in some developing nations, in the Middle East, Eastern Europe or any impoverished location with poor living conditions. Furthermore, infestations are on the rise worldwide and they are increasingly found in high-quality hotels, apartments, vehicles (cars, trains, planes, cruise ships), and even schools and hospitals. A related species, the eastern bat bug (*C. adjunctus*) occasionally enters homes when they are carried on bats that enter accessible spaces such as attics. Although this species will feed on humans, their populations are generally limited in size and they seldom reach pest status.

Waking up to bed bugs (actively feeding, crawling or mashed) is demoralising so travellers using temporary lodging (hotels, motels, hostels, resorts, etc.) or native dwellings are strongly encouraged to be observant and practise sound sanitation. Before unpacking in sleeping quarters, a through survey of the room or space should be conducted. Examine the headboard (if present), mattress (particularly the seams near the head of the bed), all linens, bed netting, bed drapes, and frame. Look for the bed bugs themselves, or dark or rusty faecal stains, pale yellow-brownish cast skins, pearly-whitish eggs (attached to any surfaces or in cracks near the droppings or cast skins) or blood spots. Also, look in/behind nightstands, books, pictures, and other wall hangings. Bed bugs can live up to a year without a blood meal and are very difficult to control. To avoid transporting bed bugs from an infested space to your personal residence, it is advisable to void unpacking near their preferred harbourages and follow advice in the 'Personal protection measures' section regarding your luggage and clothing. If bed bugs become introduced to your home, control should be conducted by a licenced pest management professional specialising in bed bug control.

Other biting Hemiptera

Many other hemipteran species are capable of biting people to some degree, although the bites are usually not severe. The exceptions include several aquatic, predatory hemipteran species that can inflict painful, defensive bites if accidentally contacted or handled. They belong to the families Belostomatidae (giant water bugs), Corixidae (water boatmen), Naucoridae (creeping water bugs) and Notonectidae (backswimmers). All these families are distributed worldwide. Although their bites generally self-resolve

Giant water bug (Belostomatidae). Source: Frank Vassen/Flickr. CC BY 2.0. Available from https://www.flickr.com/photos/42244964@N03/13569458513/

without incident, the intense stinging and numbness may last several hours, especially giant water bug and creeping water bug bites. Generally, their bites do not have any consequence of disease, but, in tropical areas of Asia, Africa and Latin America, naucorids have been implicated as mechanical vectors of Buruli ulcer (*Mycobacterium ulcerans*) although that relationship remains unclear. Use situational awareness and avoidance to prevent aquatic hemipteran bites.

Because water resources are scarce, various desert dwelling hemipterans,

Water boatman (Corixidae). Source: Peter O'Connor/Flickr. CC BY-SA-2.0. Available from https://www.flickr.com/photos/anemoneprojectors/19349498661/

Creeping water bug (*Pelicoris bipunctulus*). Source: Wolfram Sondermann/Flickr. CC BY-ND 2.0. Available from https://www.flickr.com/photos/41789001@N04/5774874898/

including plant-feeding bugs (Family Miridae), can cause an irritating but self-resolving 'pin-prick' sensation while probing perspiring people for moisture. Treatment normally is not necessary, and such bites are only a nuisance.

Backswimmer (Notonectidae). Source: Gail Hampshire/Flickr. CC BY 2.0. Available from https://www.flickr.com/photos/gails_pictures/8495923692/

Ants, bees, wasps and hornets (Order Hymenoptera)

Many ants, wasps and bees form social colonies. These social or sub-social organisations exhibit complex behaviours including defensive, stinging attacks of intruders, whether intentional or accidental. Others hymenopterans are solitary and largely non-aggressive species that use their stingers for subduing prey. Among the social groups, certain females or workers have modified, stinging ovipositors equipped with venom glands. Such stingers can inflict painful and potentially deadly reactions. The associated pain is difficult to describe, and the pain spectrum can range from irritating to excruciating. In this discussion, where available, we have provided the Schmidt Sting Pain Index as a means of assessing the relative pain levels associated with hymenopteran stings. This index ranges from Pain Level 1 (mild pain of a few minutes or less) to Pain Level 4 (blinding pain of several hours duration).

The three primary hymenopteran families responsible for most human stings are the Vespidae (wasps, hornets, yellow jackets), Apidae (honey bees, bumble bees, solitary bees), and Formicidae (ants). Wasps and ants can retract their stingers and repeatedly sting while some bees cannot. Stingers of honey bees (and those of a few tropical social wasps) are barbed so they will not be easily dislodged when they firmly penetrate skin. Once the stinger is firmly attached, the bee flies off rupturing its abdomen and leaving the poison glands pumping in venom. In addition to injecting venom, some hyme-nopterans release pheromones that serve to attract other potential related attackers.

Hymenopteran venoms are generally a complex mixture of allergenic proteins, peptides, histamines and norepinephrine, which collectively affect blood vessel relaxation and contraction. Reactions to hymenopteran stings are grouped into three broad categories based on response including: (1) immediate localised reactions; (2) systemic, toxic responses to multiple stings; and (3) systemic, allergic reactions to one or more stings. Non-allergic, localised, sting site reactions include erythema, swelling and transient pain lasting a few hours or less. Serious localised reactions may involve painful swelling of an entire extremity. Stings in the mouth or throat can result in swelling of the tongue, uvula or airway, which can close the airway resulting in respiratory distress. Systemic, toxic reactions vary from mild hives to more severe reactions such as malaise, nausea, vomiting, fever, dizziness, confusion, rash, general weakness, shortness of breath, wheezing, chest pain and even death. However, it typically takes >500 honey bee stings to get enough toxin to kill an adult human by venom toxicity alone. On the other hand, severe, allergic reactions, although rare, can be triggered by a single sting and can come on rapidly causing anaphylactic shock, breathing difficulty, and death. Of deaths attributed to hymenopteran stings, 50% of deaths occur within 30 minutes and 75% within 4 hours. These hypersensitive, systemic reactions are more severe the shorter the time interval since the previous sting. There is no known allergic cross-reactivity between honey bee and wasp venoms, although cross-reactivity may exist to some extent between different wasp venoms. Therefore, a person sensitised to one wasp species venom could potentially react to that of another species of wasp. Although rare, other reactions may include serum sickness, acute kidney inflammation and Guillain-Barré syndrome (a neurological condition). There also may be delayed reactions, which, although uncommon, include large localised swellings or systemic syndromes. These unusual responses do not always involve immunological mechanisms, but it remains unclear why these unusual responses occur.

Treatment of hymenopteran stings depends on a person's reaction. Topical corticosteroids and analgesics or systemic antihistamines are useful for treating most hymenoptera stings. However, immediate medical intervention is recommended if the victim begins showing symptoms of shock and/or anaphylaxis, or is having difficulty breathing. To the extent possible, emergency health personnel should be informed of the victim's medical history and physical condition to determine accurately the severity of the reaction. Complications may arise when hypersensitive individuals lose consciousness following a sting and cannot communicate their medical history. Therefore, anyone who has previously experienced anaphylaxis or know they are hypersensitive to hymenoptera venom should always wear a Medic-Alert tag. It is also advisable to carry a sting kit, and a personal health summary (see 'Personal protection measures'). Com-mercially available sting kits include antihistamine tablets and a preloaded epinephrine syringe (EpiPen®), or similar instrument. Additionally, hypersensitive individuals should not travel alone when hymenopteran stings are a risk. Finally, venom immunotherapy may reduce, but not eliminate, the risk of anaphylactic shock in sensitive individuals.

Several precautionary steps can be practised to minimise the risk of hymenopteran stings. For example, avoid wearing brightly coloured floral-pattern clothes or going barefoot outdoors, especially where bees and wasps are feeding close to the ground. To the extent possible and practical, avoid scented sprays, perfumes, shampoos and soaps when outdoors because they may attract bees and wasps. Additionally, exercise caution around rotting fruit, garbage cans and littered picnic grounds that tend to attract bees and wasps. Similarly, yellow jackets and bees are attracted to foods, drinks and sweets, especially sodas and fresh fruits. Finally, bees, wasps and ants are most aggressive around their nests so those should be avoided where possible.

Ants
(Family Formicidae)

Some 10 000 known species of ants inhabit various terrestrial habitats worldwide. The capacity to inflict venomous stings or produce poisonous droplets and secretions is nearly universal among ants, although most are not medically important. Ant venom is similar to other hymenopteran venoms, having neurotoxic and/or histolytic properties. A few species of ants can cause serious reactions or painful bites and stings, and they are covered here.

Bulldog ant (*Myrmecia* spp.). Source: Marshall Hedlin/ Flickr. CC BY-SA 2.0. Available from https://www.flickr. com/photos/23660854@N07/4364590418/

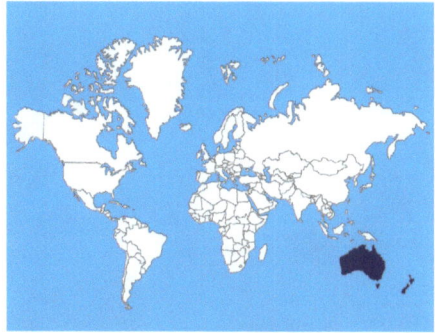

Distribution of Bulldog ants (*Myrmecia* spp.).

Bulldog and jumper ants (Myrmecia *spp.)*

Approximately 90 *Myrmecia* species inhabit sandy or mountainous areas throughout south-eastern Australia, Tasmania and New Caledonia, and they have been introduced into New Zealand. They are large, up to 25 mm (1 inch) long, aggressive ants, and they can jump up to 76 mm (3 inches), thus the name. Bulldog ants are larger and a darker colour than jumper ant. Their venom contains histamine and histamine-releasing agents, which produce excruciatingly painful stings (i.e. the black-headed bulldog ant, *M. nigriceps,* is rated at a 2 on the Schmidt Sting Pain Index). Victims also may exhibit allergic reactions, but cross-reactivity between dif-

ferent *Myrmecia* species is not well known. One species, *M. pilosula*, causes the great majority of Australia's arthropod-related allergic reactions. Although the prevalence is unknown, one study suggests roughly 50% of Australia's jumper ant allergic reactions are life threatening.

Bullet ants

Bullet ants are large, 25 mm (~1 inch) long, Central and South American ants. They can inflict excruciatingly painful stings leaving swollen, fluid-filled wounds. Rated at a 4⁺ on the Schmidt Sting Pain Index, *Paraponera clavata* is considered the most painful and crippling hymenopteran sting known. Paresthesia, vomiting, trembling and severe

Bulldog ant mandibles. Source: Stephen L. Doggett

Bullet ant (*Paraponera clavata*). Source: Bernard DuPont/Flickr. CC BY-SA 2.0. Available from https:// www.flickr.com/photos/berniedup/8430587968/

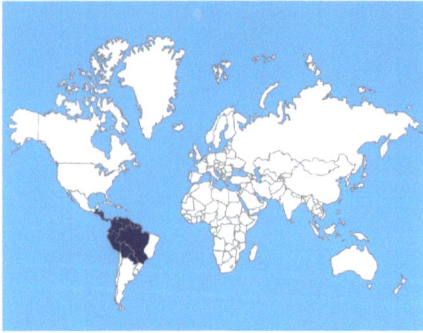

Distribution of Bullet ant (*Paraponera clavata*).

inflammation sometimes accompany the pain, which may come in waves lasting up to 24 hours.

Fire ants
(Genus Solenopsis)

Both the red imported fire ant (*Solenopsis invicta*) and black imported fire ant (*S. richteri*) typically form large colonial mounds extending above and below the ground. They also form colonies in or under protected spaces including walls, ceilings, electrical panel boxes, water meter enclosures, firewood piles, driveways and patios. Fire ants aggressively defend their nests. When disturbed, they immediately respond within seconds by climbing on any nearby object, including people, to inflict bites and painful stings rated 1 to 2 on the Schmidt Sting Pain Index. First, they use their mandibles to pinch the skin, and then they begin stinging. Each ant repeatedly stings resulting in a line or semicircular pattern of stings. In addition to the painful stings, envenomation by these ants can lead to severe, hypersensitive medical reactions. Other species of fire ants can also inflict painful stings, but these are much less common and seldom encountered.

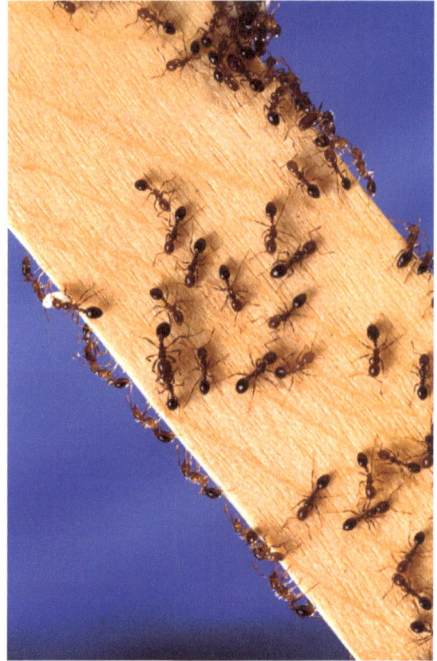

Red imported fire ant (*Solenopsis invicta*). Source: Flickr/USA Department of Agriculture. Public domain. Available from https://www.flickr.com/photos/usdagov/35630202675/

The initial fire ant sting reaction is an intense burning sensation followed shortly by formation of a wheal up to 13 mm (0.5 inch) in diameter, and itching and swelling around the sting site. Approximately 4

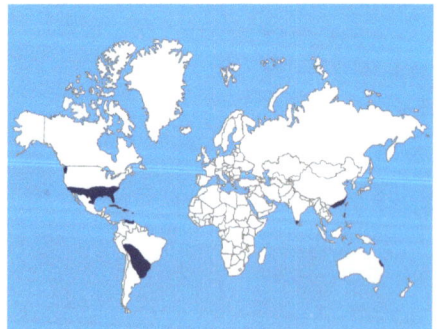

Distribution of red fire ant (*Solenopsis invicta*).

Distribution of black fire ant (*Solenopsis richteri*).

Floating fire ants on floodwater. Source: US National Park Service

Fire ant mound showing agitated ants. Source: David E. Bowles

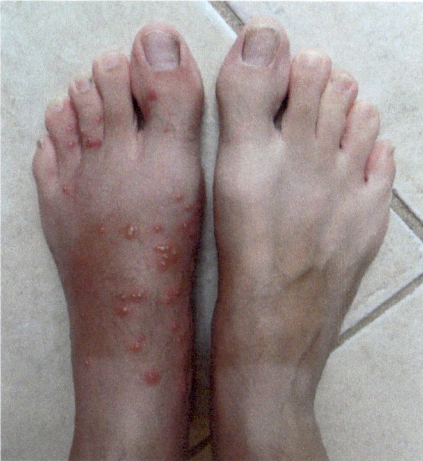

Fire ant bite pustules on foot. Source: Bart Everson/ Flickr. CC BY 2.0. Available from https://www.flickr.com/photos/editor/4945179598/

hours later, fluid-filled vesicles begin forming, which become white necrotic lesions or pustules after roughly 24 hours. Pustules may last a week or more and once broken they itch intensely. Each lesion heals leaving a small scar.

When in the range of fire ants, take care to avoid fire ant mounds, and do not venture outdoors wearing open toed footwear or barefoot. Wearing permethrin treated, long pants, tucking the pants into your socks or boots may offer some protection. When attacked by fire ants, the hands or a cloth should be used to remove them from the body as quickly as possible. Topical corticosteroids, systemic antihistamines and, to a limited extent, hot showers or baths serve to relieve itching. Avoid scratching pustules to prevent secondary infection.

Harvester ants
(Genus Pogonomyrmex)

Harvester ants (*Pogonomyrmex* spp.) are large, relatively gentle ants that will rapidly defend the area surrounding their mounds. When provoked or handled, they will sting. The response to the sting may be slow, and it may take up to 30 seconds to feel the pain. The pain from a harvester ant

Harvester ant (*Pogonomyrmex maricopa*). Source: Brett Morgan/Flickr. Available from https://www.flickr.com/photos/131104726@N02/31538269553/

Harvester ant mound. Source: David E. Bowles

builds to an excruciating intensity. For example, the sting of the red harvester ant, *P. barbatus,* is rated 3 on the Schmidt Sting Pain Index, and it may last several hours. These ants are common in arid portions of the United States southward to South America. First aid measures described for

Distribution of harvester ants (*Pogonomyrmex* spp.).

fire ants and cold compresses may be useful for relieving pain from these stings.

Other ants

Many other ants can sting, but the majority only cause localised reactions and annoyance. However, the stings of some species occasionally may cause allergic reactions. For example, the Middle Eastern samsum ant (*Pachycondyla sennaarensis*) has caused serious systemic reactions in people. Similarly, pavement ants (*Tetramorium caespitum),* a native European species, were introduced to the United States and are now established in many urban areas. Although the stinger of this species has difficulty piercing human skin, it can cause temporary, intense itching. Following multiple stings, reddened spots may appear within a few days, leading to chronic itching. The trap jaw ants (genus *Odontomachus*) found in Central to South America, Asia, Australia, Africa and the south-eastern United States have a sting that causes an immediate burning sensation, but this sensation rapidly dissipates. However, there have been cases of allergic reactions in some people that resulted in swelling of the face and extremities, and associated itching, all of which may last up to a week. In addition, it is thought that some protein-feeding ants such as Pharaoh ants (*Monomorium pharaonis*) mechanically transit pathogens in hospitals. Finally, despite the mass media portrayal, army ants are largely harmless to humans, but the large workers (major workers) have formidable mandibles that can inflict a painful defensive bite. The common name army ant has been applied to many different species of ants worldwide of which the genera *Eciton* and *Dorylus* are representative.

Army ant (*Dorylus* sp.) head showing mandibles. Source: Bernard DuPont/Flickr. CC BY-SA 2.0. Available from https://www.flickr.com/photos/berniedup/7073859635/

Bees
(Superfamily Apiodea)

The bees constitute a speciose (>20 000 species) group of insects belonging to several families. Many are ecologically important pollinators that are critical in the maintenance of certain agricultural crops, and they can serve as a source of food (e.g. honey). Bees may be either solitary or social in behaviour. The majority of bees pose no threat to people, but a few can inflict painful stings that can result in serious medical complications.

Honey bees
(Apis *spp.)*

European or western honey bees (*Apis mellifera*) are the primary domesticated bees worldwide, but other species are distributed primarily in Asia. Honey bees can be found worldwide except the northernmost boreal areas. These highly social insects are typically non-aggressive and their pollination services, honey, and other products greatly benefit humanity. Honey bees do present a health threat to people and pets if they are provoked or accidentally contacted, because they can inflict a painful sting. The

sting is rated 2 on the Schmidt Sting Pain Index. The honey bee's stinger is barbed so when it enters the skin and the bee flies away, the stinger and venom gland remain behind and continue injecting venom into the victim. Individual honey bees are not considered dangerous to people who are not allergic to the venom. However, swarming attacks that produce multiple stings and allergic, anaphylactic shock reactions to even a single sting can cause severe injury and death unless rapidly and properly treated (see reactions and treatments presented earlier). Africanised bees, or 'killer bees', now widely distributed in the southern United States and southward to include most of South America, are a more aggressive, extremely dangerous genetic strain of honey bees that fiercely attack any perceived threat. Although typical honey bees also may attack in mass when threatened, Africanised bees are much more likely to demonstrate this behaviour. Attacks by Africanised honey bees may result in literally hundreds of stings.

Attacking bees can be very persistent and may pursue their victim over distances up to 3 km (2 miles). If attacked by bees, the best course of action is to get inside a house or other safe building, vehicle or other closed space. If an enclosed space is not available, the alternative is to run in zig-zag fashion, if possible, through brushy or wooded areas. This helps break-up their search image making it easier to elude them. Avoid entering water such as streams and ponds because bees may continue to search the area up to an hour, longer than the average person can swim or hold their breath under water.

Because the honey bee's barbed stinger remains in the skin injecting venom and

Honey bee (*Apis mellifera*) worker. Photo: Harold J. Harlan

Honey bee swarm. Source: US National Park Service

releasing pheromone, they should be removed from the skin as quickly as possible, especially in multiple sting attacks. Do not attempt to remove the stinger with the fingers because the pinch compresses the venom glands, forcing venom into the wound. In most cases, the best means of removing stingers is to use the most readily available, hard, straight edged instrument or tool (e.g. credit card, knife edge) to scrape out the stinger(s). This is especially important when there are multiple stings. Once the stinger is removed, clean the area around the sting to neutralise the pheromones that may be present. This is particularly important if you must be outside and bees are still

Honey bees on comb. Source: Andria/Flickr. CC BY-ND-2.0. Available from https://www.flickr.com/photos/fauxlaroid/9523401087/

present. Any compound that denatures protein works including alcohol, sting swabs, or soap and water. Normally, systemic antihistamines, cold compresses, and aluminium sulfate (Stingose®) are effective for treating reactions to stings, but multiple stings and allergic reactions may require emergency medical treatment. People with known hypersensitivities to honey bee venom should always carry a sting kit with an EpiPen®, or similar device while outdoors.

Bumble bees and carpenter bees

Bumble bees (*Bombus* spp.) and carpenter bees (*Xylocopa* spp.) have a similar appearance, but bumble bee abdomens are hairy while carpenter bee abdomens are smooth and shiny. Both groups of bees have a broad global distribution. Bumble bees live in small colonies while carpenter bees are solitary. Bumble bees and carpenter bees are normally passive, but will sting in defence or if accidentally contacted, and the stings are painful (rated at a 2 on the Schmidt Sting Pain Index). Reactions to the stings of these bees are similar to other

Bumble bee (*Bombus* sp.). Source: Mike Bowler/ Flickr. CC BY 2.0. Available from https://www.flickr.com/photos/mbowler/530339037/

Carpenter bee (*Xylocopa virginica*). Source: Judy Gallagher/Flickr. CC BY 2.0. Available from https://www.flickr.com/photos/52450054@N04/36389208592

Distribution of bumble bees (*Bombus* spp.).

bees, including anaphylaxis (but this occurs much less frequently in comparison with honey bee stings).

A solitary bee (Family Andrenidae). Source: Janet Graham/Flickr. CC BY 2.0. Available from https://www.flickr.com/photos/149164524@N06/33439491774/

Other bees

There are many other types of bees distributed worldwide. Most of these bees are solitary, non-aggressive, and they rarely sting people. Although stings do occur, they are only mildly painful (i.e. a pinprick) and generally are not consequential.

Wasps and hornets
(Family Vespidae)

All true wasps can sting, but most are solitary and normally use their sting to kill or paralyse prey. In comparison, vespid wasps, most notably the hornets (*Dolichovespula* spp. and *Vespa* spp.), yellow jackets (*Vespula* spp.) and paper wasps (*Polistes* spp.) are social wasps. This large family includes over 5000 species distributed worldwide. They form various sized colonies and aggressively defend their nests when disturbed or threatened. Most vespids have various black, yellow, reddish or whitish markings. They use wood particles, foliage and saliva to construct their paper nests. Social wasps select protected areas including underground spaces (yellow jackets) or sites such as tree limbs and tree cavities, or sheltered areas such as

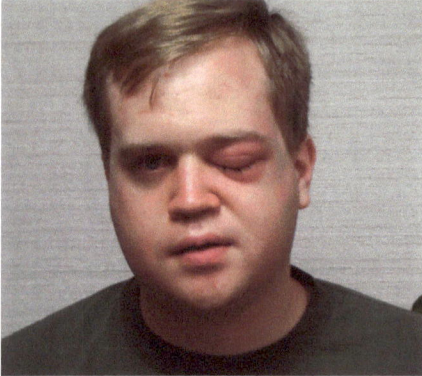

Response following a yellow jacket sting to the eye lid. Source: James A. Swaby

Bald-faced hornet (*Dolichovespula maculata*). Source: Judy Gallagher/Flickr. CC BY 2.0. Available from https://www.flickr.com/photos/52450054@ N04/20779113131/ -eHuHF4-AMYmM4-oPrPKw-PyWT3v-q617Pw-fLXvwG-6EgRZn-4KtmwH

roof gables (hornets, paper wasps) to construct their nests.

Stings by these wasps are painful. For example, stings by yellow jackets and bald-faced hornets are rated 2 on the Schmidt Pain Index, while the paper wasp is rated 3. In addition to the intense pain they cause, stings may cause localised swelling, but they rarely produce significant reactions such as anaphylaxis. One notable exception is the Asian giant hornet, *Vespa mandarinia*. During 2013, the Asian giant hornets killed 41 people and injured more than 1600 in Shaanxi Province, China. Throughout their range, this species is responsible for multiple deaths annually. Most deaths attributed to this species are due to anaphylactic shock, cardiac arrest and renal failure. Multiple stings from this species usually require emergency medical treatment. Some victims of Asian giant hornet stings have experienced multiple organ failure, tissue necrosis and haemorrhaging. Other aggressive hornets with potentially dangerous stings include the European hornet (*Vespa crabro*), oriental hornet (*Vespa orientalis*) and Asian predatory wasp (*Vespa velutina*), all of which have been widely introduced outside of their native range to other areas of the world.

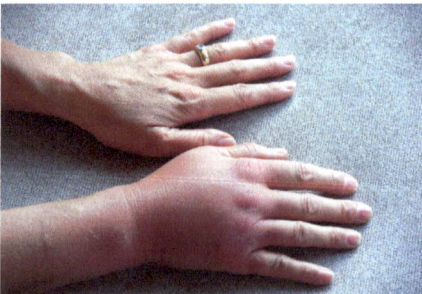

Oedema of the hand following a yellow jacket sting. Source: Stephanie Young Merzel/Flickr. CC BY 2.0. Available from https://www.flickr.com/photos/justthismoment/2693328754/

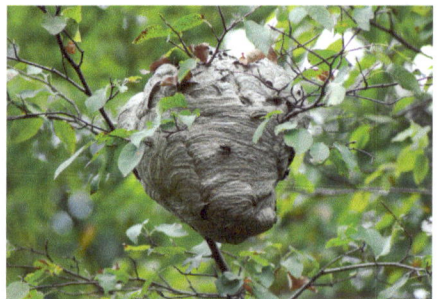

Nest of bald-faced hornet. Source: BobMacInnes/ Flickr. CC BY 2.0. Available from https://www.flickr.com/photos/lonetown/1352052086/

Yellow jacket (*Vespula* sp.). Source: David Hill/Flickr. CC BY 2.0. Available from https://www.flickr.com/photos/dehill/10860353154/

Yellow jacket nest. Source: Donald Hobern/Flickr. CC BY 2.0. Available from https://www.flickr.com/photos/dhobern/14599730913/

Paper wasps (*Polistes* sp.) guarding their nest. Source: Ian Sane/Flickr. CC By 2.0. Available from https://www.flickr.com/photos/31246066@N04/4936872846/

For most vespid stings, systemic antihistamines and cold compresses are effective for treating sting reactions, but allergic reactions may require emergency medical

European hornet (*Vespa crabro*). Source: Judy Gallagher/Flickr. CC BY 2.0. Available from: https://www.flickr.com/photos/52450054@N04/5911185508/

treatment. In most cases, intense pain dissipates within an hour or so, but the surrounding tissue may be tender and sensitive for a couple of days. Unlike honey bees, individual wasps and hornets can sting multiple times. Similar to honey bee attacks, the best course of action when attacked by social wasps is to get inside a house or other safe building, vehicle or other closed space. If an enclosed space is not available, running through brushy or wooded areas may be an effective alternative. In the case of multiple stings, or if systemic symptoms are exhibited (actual or suspected), emergency

Asian giant hornet (*Vespa mandarinia*). Source: Kenpai/Wikipedia. CC BY-SA 3.0. Available from https://en.wikipedia.org/wiki/Asian_giant_hornet#/media/File:Vespa_mandarinia_japonica1.jpg

Oriental hornet (*Vespa orientalis*). Source: S. Rae/ Flickr. CC BY 2.0. Available from https://www.flickr. com/photos/35142635@N05/8703295920/

Asian predatory wasp (*Vespa velutina*). Source: Gilles San Martin/Flickr. CC BY-SA 2.0. Available from https:// www.flickr.com/photos/sanmartin/33965402201/

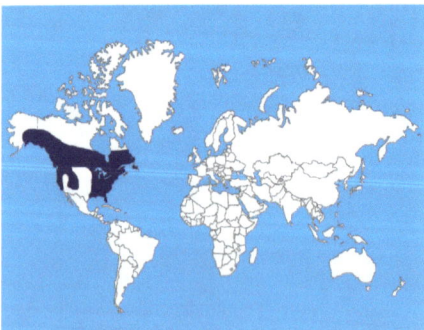

Distribution of bald-faced hornet (*Dolichovespula maculata*).

Distribution of European hornet (*Vespa crabro*).

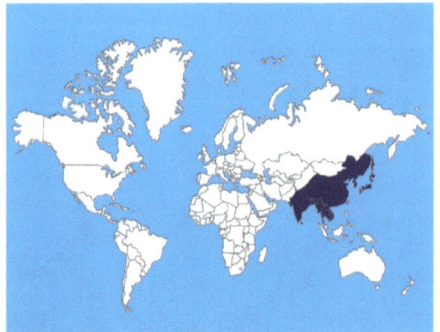

Distribution of Asian giant hornet (*Vespa mandarinia*).

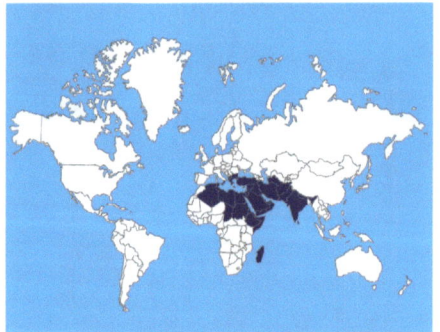

Distribution of oriental hornet (*Vespa orientalis*).

medical attention or treatment should be sought immediately. People with known hypersensitivity to wasp stings should always carry a sting kit with an EpiPen®, or similar device while outdoors.

Velvet ant (species unidentified). Source: Judy Gallagher/Flickr. CC BY 2.0. Available from https://www.flickr.com/photos/52450054@N04/27111924110/

Distribution of Asian predatory wasp (*Vespa velutina*).

Velvet ants
(Family Mutillidae)

There are ~8000 species in this family that are distributed worldwide. Velvet ants are not ants, but rather wingless, female wasps. The males of all velvet ant species are winged and not capable of stinging. The dense setae covering their bodies and red or orange markings give them a hairy, rather attractive, velvety appearance. Some species can be large reaching over 25 mm (1 inch) in length. They are normally not aggressive, but will readily sting when threatened (i.e. by handling or accidental contact). Because they are often strikingly beautiful in colouration, children occasionally attempt to handle them, which can

result in stings. The sting of velvet ants is excruciatingly painful, rated at a 3 on the Schmidt Sting Pain Index. Due to their large size and painful sting, they sometimes are called 'cow killers.' Simple avoidance and using personal protection measures are the best means to prevent wasp stings.

Spider-hunting wasps
(Family Pompilidae)

These large, 76 mm (3 inch), long wasps have distinctively marked orange-yellow coloured wings and iridescent purple-blue bodies. They are broadly distributed over much of the Southern Hemisphere. Spider-hunting wasps are not aggressive and

Velvet ant (*Dasymutilla occidentalis*). Source: Harold J. Harlan

Tarantula hawk wasp (*Pepsis grossa*). Source: Ken Bosma/Flickr. CC BY 2.0. Available from https://www.flickr.com/photos/kretyen/2988343797/

Tarantula hawk wasp preying on a tarantula. Source: David Crummey/Flickr. CC BY 2.0. Available from https://www.flickr.com/photos/dcrummey/5923526296/

Mud dauber (*Chalybion* sp.). Source: Urasimaru/Flickr. CC BY-SA 2.0. Available from https://www.flickr.com/photos/urasimaru/9158791516/

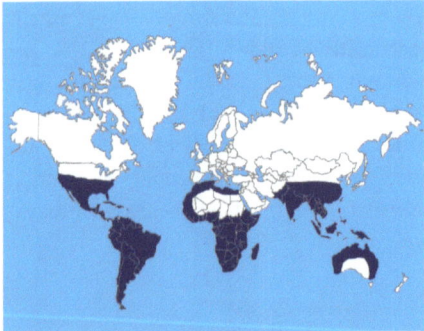

Distribution of tarantula hawk wasps (*Pepsis* spp., *Hemipepsis* spp.).

Mud dauber (*Sceliphron caementarium*). Source: David Hill/Flickr. CC BY 2.0. Available from https://www.flickr.com/photos/dehill/9587575810/

rarely sting unless provoked or handled. Their stings are excruciatingly painful. For example, tarantula hawk wasps (genera *Pepsis* and *Hemipepsis*) are considered to have one of the world's most painful stings rated at a 4 on the Schmidt Sting Pain Index, and they are second only to the bullet ant in that regard. The adult female wasps attack tarantulas by depositing eggs on them that subsequently hatch into larvae and feed on the tarantula. Spider-hunting wasps are distributed worldwide.

Cicada-killer (*Sphecius speciosus*). Source: David E. Bowles and Mark Pomerinke, United States Air Force

Other wasps

Other hymenopteran families contain non-aggressive wasps that either do not sting or whose sting only produces mild, inconsequential pain. Notable among these are the sphecid wasps (Family Sphe-

ANTS, BEES, WASPS AND HORNETS

cidae) that include the brightly coloured mud daubers (*Sceliphron* spp., *Chalybron* spp., *Trypoxylon* spp.), which construct variously shaped mud nests. Because they are often located in association with houses and other buildings, they commonly are encountered and may be mistaken for social wasps. Large-bodied cicada killers (*Sphecius* spp.) are distrib- uted worldwide and have an ominous appearance, especially when they aggre- gate in swarms for mating. However, these wasps are not aggressive at all and are gen- erally reluctant to sting. The few reports of their stings indicate they are only mildly painful and not unlike a pin-prick. Cicada killers are often commonly encountered in urban settings.

Moths
(Order Lepidoptera)

Of the 300 000 species of moths and butterfly distributed worldwide, ~100 can cause severe human reactions. They belong to the families Limacodidae, Lymantriidae, Megalopygidae, Saturniidae, and Thaumetopoeidae. The caterpillars of these moths generally have a clearly spiny or soft fuzzy appearance (see photos). Many harmless caterpillars, such as woolly bears, may resemble urticating caterpillars, but they are harmless. Urticating moths and caterpillars inhabit temperate and tropical environments worldwide. Although the larval stage or caterpillar is typically the life stage implicated in envenomation, all life stages can have hollow, urticating bristles (setae) that easily break off, penetrate the skin, and release a venomous combination of proteins, enzymes and histamines. Individual reactions to this urtication depend on the part of the body contacted, duration of the contact, size of the insect and sensitivities to the chemicals. This action of stinging and urtication is often termed lepidopterism and, when larvae are solely involved, erucism may be used. There are two broad groups of lepidopterism based on reaction: (1) reactions to adults, eggs or pupal cocoons that produce irritating contact dermatitis, and (2) reactions to larvae and adults that can range from a mild burning sensation to extreme pain lasting up to 12 hours, followed by residual pain up to 2 weeks. Following the initial pain of a sting, additional reactions may include a rash at the sting site, itching, reddened skin, reddened wheals, dermatitis, lesions, localised swelling and formation of blisters or pustules. Contact with the eyes may cause painful conjunctivitis. More serious responses may include red blood cell destruction, haemorrhage, tissue necrosis, inflamed nasal membranes, sinuses and lung tissue, which may obstruct breathing. Other serious systemic complications include swollen lymph nodes of the underarm and groin, headache, nausea, fever, difficulty breathing and prostration.

When treating urticating wounds, the care provider should wear protective gloves and use adhesive tape to strip the setae from the wound site first. Care should be

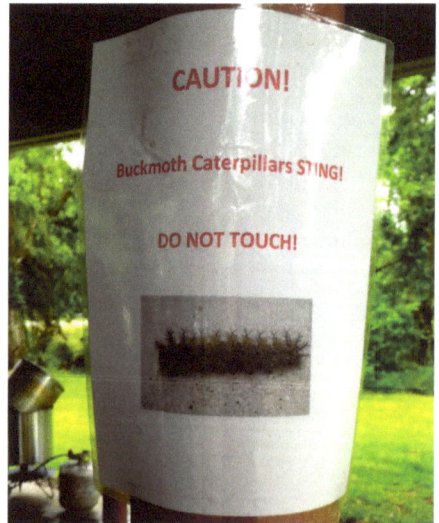

Warning sign for the presence of stinging buck moth caterpillars. Source: David E. Bowles

Stinging rose caterpillar (*Parasa indetermina*). Source: David E. Bowles

Brown puss caterpillar (*Megalopyge opercularis*). Source: David E. Bowles

used so the remaining setae will not be reapplied elsewhere, including on the care provider. Following removal of setae, soap and water can be used to clean the area. Ice packs are helpful to minimise swelling, but serious cases may require use of systemic corticosteroids and oral antihistamines. Severe pain can be treated with analgesics or aluminium sulfate (Stingose®). A simple paste made from baking soda and water can be used to manage pain from less painful stings. Immediate medical treatment should be sought for dangerous symptoms including, prostration, shock, difficulty breathing and debilitating, severe pain, and bleeding syndrome

(haemorrhagic diathesis or inhibition of blood coagulation). Symptoms of bleeding syndrome may include body-wide haematoma, nasal haemorrhaging, and other bleeding. The specific physiological mechanism of this syndrome is not known.

Most stings from urticating moths or caterpillars, although painful, are generally quickly resolved and of modest consequence. In contrast, the stings or reactions to a few species can present injury that is more significant and they are presented in more detail here. The South American moth genus *Lonomia* contains the most dangerous caterpillars in the World. The caterpillars *Lonomia achelous* and

Saddleback caterpillar (*Acharia stimulea*). Source: Katja Schultz/Flickr. CC BY 2.0. Available from https://www.flickr.com/photos/treegrow/37067603750/

Gray puss caterpillar (*Megalopyge opercularis*). Source: David E. Bowles

Buck moth caterpillar (*Hemileuca maia*). Source: David E. Bowles

Stinging caterpillar (Family Limacodidae). Source: hspauldi/Flickr. CC BY-SA-2.0. Available from https://www.flickr.com/photos/hspauldi/3178208721/

Lonomia obliqua (Venezuela and Brazil, respectively) inflict stings that cause bleeding syndrome. Following initial contact with these caterpillars, a burning pain occurs followed by reddened skin, a heat sensation, swelling, blisters, headache and vomiting, then the bleeding syndrome begins within ~12 hours. Contact with these caterpillars has reportedly caused deaths of adult humans.

The puss caterpillars or stinging asps (*Megalopyge* spp., southern flannel moth) are another dangerous group that has urticating caterpillars. These caterpillars inflict excruciatingly painful stings that can produce haemorrhagic lesions, significant swelling, swollen lymph nodes low blood pressure, shock and complete prostration.

Stinging asps are slow moving caterpillars and are often accidently contacted, and, because of their fuzzy appearance, children sometimes attempt to play with them. They are most commonly encountered in late summer and early fall (autumn), and often near oak trees. Stinging asps are broadly distributed from the southern United States southward throughout Central America.

Adults and larvae of certain brown-tail moths, tussock moth and *Hylesia* spp. possess urticating setae. Caterpillars of the African genus *Anaphae* (Thaumeto-

Stinging caterpillar (Family Limacodidae). Source: Bernard DuPont/Flickr. CC BY-SA 3.0. Available from https://www.flickr.com/photos/berniedup/7656548722/

Lonomia obliqua. Source: Rodrigo Morante/Wikipedia. Public domain. Available from https://en.wikipedia.org/wiki/Lonomia#/media/File:Taturana.JPG.

Distribution of *Lonomia* spp.

Adult gypsy moth (*Lymantria dispar*) and eggs. Source: Wisconsin Department of Natural Resources/Flickr. CC BY-ND 2.0. https://www.flickr.com/photos/widnr/6522464529/

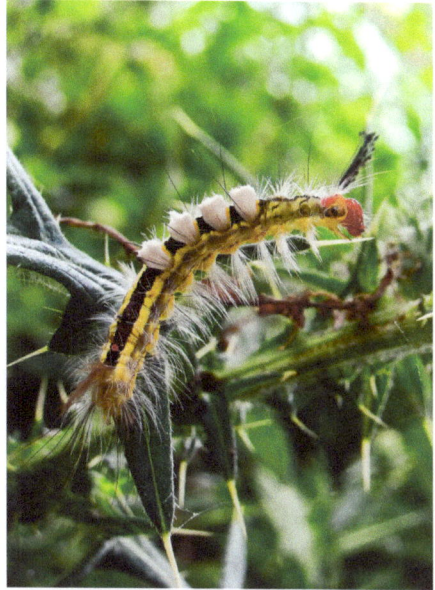

Tussock moth caterpillar (*Orgyia leucostigma*). Source: Noir imp/Flickr. CC BY 2.0. Available from https://www.flickr.com/photos/golo_ undertow/4911638659/

poeidae) and South American genus *Hylesia* (Saturniidae) setae can cause substantial skin eruptions among the human population. Similarly, tussock moths (Family Lymantridae) are widely distributed in North through South America, Europe, Japan, and South-East Asia. These caterpillars have caused large outbreaks of dermatitis among people when their populations are high during major defoliation events. Such dermatitis may be due to direct contact with the caterpillars or the lungs may become injured from inhaling the loose urticating setae falling from the forest canopy. The resulting inflammatory respiratory syndrome is called tussockosis, but it is relatively rare. This condition may become complicated when certain fungi have contaminated the setae resulting in secondary respiratory infections.

Beetles
(Order Coleoptera)

This is the largest insect order with upwards of 400 000 species distributed worldwide. Most beetles are not medically important, but some species in three families may cause medical responses in some people. These are the blister, rove and dermestid beetles.

Blister beetles
(Family Meloidae)

There are 2500 blister beetle species distributed worldwide. They range in size up to 32 mm (1.25 inches) long, are often brightly coloured or striped, and the head and thorax are narrower than the abdomen. The haemolymph (blood) of many blister beetles contains cantharidin, which is exuded when the beetle is threatened or accidentally pressed, rubbed or squashed. Wherever cantharidin touches the skin, it results in burning and tingling followed by formation of fluid-filled vesicles or blisters. Blisters are painless, but if they rupture, they can spread the cantharidin producing additional blisters.

Blister beetle (*Meloe impressus*) exuding cantharidin. Source: Allison Carey/Flickr. CC BY 2.0. Available from https://www.flickr.com/photos/allisons_photos/5605350438/

Extensive contact with blister beetles can cause complications including inflammation of the mouth and throat, excessive salivation, vomiting of blood, abdominal pain, diarrhoea and painful urination.

Blister beetle (*Meloe impressus*). Source: Harold J. Harlan

Striped blister beetle (*Pyrota bilineata*). Source Sarah Zukoff/Flickr. CC BY 2.0. https://www.flickr.com/photos/entogirl/9596123058/

Spanish-fly (*Lytta vesicatoria*). Source: Udo Schmidt/ Flickr. CC BY-SA 2.0. Available from https://www. flickr.com/photos/coleoptera-us/32966331712/

Blister beetle (*Epicauta waterhousei*). Source: LiCheng Shih/Flickr. CC BY 2.0. Available from https://www.flickr.com/photos/papilioshih/ 34812300760/

The Spanish fly (*Lytta vesicatoria*) has been popularised as an aphrodisiac, but this notion is incorrect and consumption of these insects or their extracts can cause painful injury. Most blisters require only first aid and will self-resolve. Diluting the toxin with soap and water, isopropyl alcohol or other suitable disinfectant helps minimise additional blistering. Wounds should be treated with a topical corticosteroid or antibacterial salve. Scratching the wounds should be avoided to prevent secondary infections. Severe reactions require emergency medical care.

Rove beetles
(Family Staphylinidae)
About 63 000 species (estimates vary) belong to this large beetle family. Rove beetles are characterised by elongated, slender bodies and short forewings (elytra). Among the rove beetles, only members of the genus *Paederus* present a potential medical threat to people. Over 600 species in this genus are distributed worldwide. *Paederus* are relatively small beetles, being less than 7 mm (0.3 inch) in length. The haemolymph of *Paederus* contains the world's most potent contact toxin, a compound named pederin ($C_{24}H_{43}O_9N$), which is purported to be 12 times more toxic than cobra venom. The beetles do not bite or sting. Rather, exposure occurs when the beetles are accidently crushed against the skin. Contact with the beetles primarily occurs when the beetles are attracted to light where people are located. Contact with the haemolymph can cause severe dermatitis, including blistering, pustule formation and conjunctivitis of the eyes. Initial symptoms following contact include reddening of the skin, and a burning sensation. Painful irritation and itching follows along with extensive formation of pustules and blisters within ~4 days following the exposure. Blisters can become quite large approaching 15 mm (0.6 inch) in diameter. The blistering commonly follows a linear pattern indicating the track the beetle travelled on the skin. Recovery generally occurs 7–10 days following exposure and normally without any long-term consequences. Exposure to direct sunlight may lead to both hyper- and hypo-pigmentation (i.e. a dark scarring) in some people.

Cases of dermatitis due to *Paederus* have been reported from numerous locations around the world. Large outbreaks (≤200 cases) have been reported from central and northern Africa, south and south-east Asia, north-eastern Australia, south-western United States and Central South America. The Asian coastal species,

Rove beetle (*Paderus* sp.). Source: Stephen L. Doggett

Khapra beetle adult (*Trogaderma granarium*). Source: Udo Schmidt/Flickr. CC BY-SA 2.0. Available from https://www.flickr.com/photos/coleoptera-us/30364726883/

Paederus fuscipes, is particularly harmful, and has been responsible for large outbreaks in some areas, including thousands of cases in Malaysia.

Situational awareness and avoidance is key to preventing contact with rove beetles. Wear protective clothing, including long sleeves and pants, to minimise exposure of the beetles to skin. Turn off or minimise lighting at night and avoid sitting near lights that may attract the beetles. In lodging, or tents if camping, keep doors/windows/tent flaps closed, and use a bed net if practical. If exposure does occur, wash the affected areas with soap and water as soon as possible. Symptoms can be alleviated by using cold compresses, calamine lotion, aloe vera, and topical corticosteroids or antihistamines. For severe reactions, or if the eyes are involved, professional medical attention should be obtained.

Dermestids
(Family Dermestidae)
These beetles can contaminate stored dry goods (grains, cereals, nuts, etc.) and other materials (books, paper, cardboard, taxidermy mounts, etc.) causing economic losses. They can also potentially threaten human health. Dermestid larvae shed barbed hairs (setae) that sometimes cause serious allergic reactions in some individuals. The barbed setae can also penetrate the gastrointestinal lining and form cysts. Although these problems can be caused by a variety of dermestid species, the Khapra beetle (*Trogaderma granarium*) is one of the most commonly implicated species. Situational awareness and avoidance are the best means of avoiding problems with dermestids. Always inspect dry foods for the presence of dermestids and other pest insects, and do not consume those products.

Khapra beetle larvae (*Trogaderma granarium*). Source: United States Customs and Border Protection

Flies
(Order Diptera)

Flies cause considerable human suffering throughout the world and they are considered to be the largest group of medically important invertebrates. Of the 150 000 known fly species, several thousand species have been implicated as pests or vectors of disease. Flies go through complete metamorphosis with eggs, larvae, pupae and adults, and the larval and adult stages cause the medically related or nuisance problems. Three types of medically important flies are considered here, including: (1) myiasis flies; (2) biting flies; and (3) filth flies.

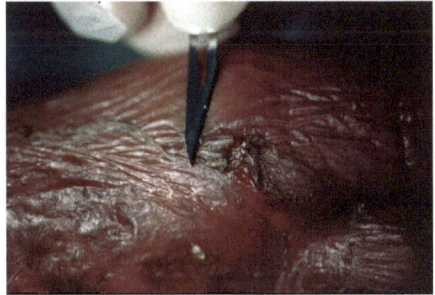

Non-specific myiasis in the wound of a human patient. The white maggots can be seen to the right of the scalpel. Source: Dr Holland, United States Navy. Courtesy of Dr Rapini

Typical life stages for a fly: larva (maggot), pupa, adult. Source: Martin Cooper/Flickr. CC BY 2.0. Available from https://www.flickr.com/photos/m-a-r-t-i-n/16544156480/

Myiasis flies

Many flies can cause human myiasis, but only a few are frequent or severe enough to merit inclusion here. For example, at least 50 species belonging to the families Muscidae, Calliphoridae and Sarcophagi-

dae can contaminate uncooked food when they visit it causing psychologically traumatic enteric or intestinal pseudomyiasis. By comparison, other flies that normally breed in meat, carrion or living tissue can invade human flesh causing true myiasis. Here, we cover the more important ones, treatment, and prevention.

Human bot fly
(Family Cuterbridae)

Human bot flies (*Dermatobia hominis*) or torsalo appear superficially similar to large bees, and they are found from south-eastern Mexico, southward throughout much of South America. Bot flies parasitise a broad range of mammals, including humans. Adult bot flies lack mouthparts and do not bite or lay eggs directly on the host. Rather, they glue eggs onto ticks or biting flies such as mosquitoes, which then transport the eggs to the potential host

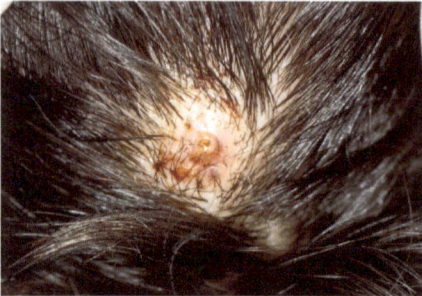

A bot fly (*Dermatobia hominis*) buried in the scalp of a human patient: Source: Dr Holland, United States Navy. Courtesy of Dr Rapini

A bot fly larva. Source: David E. Bowles

where they hatch. Upon hatching, the larvae quickly emerge and penetrate the skin to feed in a dermal pocket over a period of 5 to 12 weeks. The feeding action produces a swollen, painful, itchy, warble-like lesion in which the larvae can be felt

moving around. The infestation is self-limiting, ending when the final, rather large, larval stage exits the dermal pouch and falls off to pupate in the soil. See treatment and prevention below.

Tumbu and Lund's flies (Family Calliphoridae)

Tumbu or mango fly (*Cordylobia anthropophaga*) and less often, Lund's fly (*Cordylobia rodhaini*) cause human myiasis. The tumbu fly occurs over much of tropical sub-Saharan Africa and south-western Saudi Arabia. By comparison, Lund's fly inhabits tropical rainforest from Senegal to Central Africa and south to Angola and Zimbabwe. Several wild rodent species are the preferred hosts for both species, and dogs are common domestic hosts. Both species deposit their eggs below the surface of sandy soil and occasionally on clothing tainted with faeces or urine. The larvae respond to any disturbance of the soil and rapidly migrate to the surface where they climb on the host and penetrate the skin. They burrow into subcutaneous tissue where they feed for 10 to 12 days, forming a painful and itchy boil-like lesion. When feeding and larval growth are complete, they exit the hosts' tissues, fall to the ground and

Distribution of Human bot fly (*Dermatobia hominis*).

Tumbu fly larva (*Cordylobia anthrophaga*). Source. Stephen L. Doggett

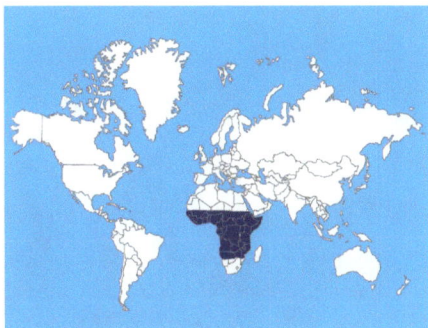

Distribution of tumbu fly (*Cordylobia anthropophaga*).

Mud hut in African village. Tumbu fly commonly inhabit the roofs and floors of these huts. Source: John Atherton/Flickr. CC BY 2.0. Available from https://www.flickr.com/photos/gbaku/3267834892/

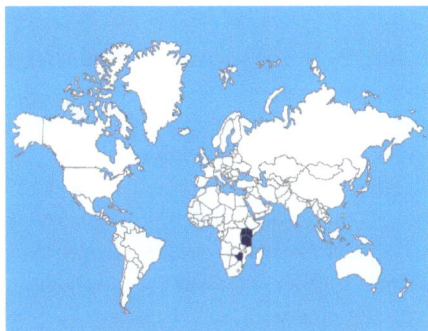

Distribution of Lund's fly (*Cordylobia rodhaini*).

pupate in the soil. See also 'Myiasis treatment and prevention'.

Congo floor maggot (Family Calliphoridae)

Congo floor maggot (*Auchmeromyia senegalensis*) occurs throughout tropical sub-Saharan Africa and the Cape Verde Islands. Burrow-dwelling mammals are their primary hosts, but they also invade mud huts where they bite humans. The orange coloured, stout bodied adult flies are difficult to see among roofing materials where they hide. Females emerge to lay eggs in the mud floors, especially urine-contaminated areas or under floor coverings. The nocturnal larvae then attack people while they sleep. They do not climb, so exposure occurs when the skin makes direct contact with the infested soil. The final instar maggots are large up to 18 mm (0.7 inch) in length. Once on the host, their hooked mouthparts latch on to the skin, where they slice down to a blood vessel, and they use sucking mouthparts to feed on the released blood. After feeding for 15 to 20 minutes, they leave the host and retreat back to cracks under sleeping mats or burrow into the soil until their next blood meal. Feeding may occur daily until larval development is complete. Their

Distribution of Congo floor maggot (*Auchmeromyia senegalensis*).

entire life cycle lasts 10 weeks, and they breed year-round. Following feeding, the victim may develop atypical myiasis lesions that swell and itch, although they are not particularly painful. See also 'Myiasis treatment and prevention'.

Sheep bot fly
(Family Oestridae)

Ocular myiasis due to sheep bot fly (*Oestrus ovis*) is relatively common world-wide wherever sheep are raised including around the Mediterranean, Middle East, Africa, Azerbaijan, Central America and South America. A few cases have been reported from North America. The larvae are obligate parasites of sheep, goats, camels and horses, infesting their nostrils and frontal sinuses, but humans occasionally become infested as well. Adult females, deposit immatures already developed into larvae (larviposit) directly onto the eye. This produces a painful but generally self-resolving conjunctivitis. Sometimes larvae penetrate the inner eye, causing serious complications and requiring medical treatment. See also 'Myiasis treatment and prevention'.

Sheep maggots
(Family Calliphoridae)

About a dozen species of flies (*Chrysomya* spp.) known as 'sheep maggots' are confined to Old World tropical and semi-tropical regions (i.e. Cameroon, Congo, Ethiopia, Kenya, Lesotho, South Africa, Tanzania and Zimbabwe). One species, *C. chloropyga*, is commonly implicated in human myiasis in South Africa and surrounding areas. This species has been introduced throughout the Americas and the Caribbean. Females lay eggs singly or

Old World blow fly (*Chrysomya megacephala*). Source: Muhammad Mahdi Karim/Wikipedia. GFDL. Available from https://en.wikipedia.org/wiki/Chrysomya#/media/File:Ch.megacephala_wiki.jpg

in batches inside wounds, sometimes on unbroken skin covering bruises or abscesses, and occasionally on soiled blood spots. Young larvae feed for about a day on liquids exuding from such tissues, then invade and feed upon living tissue. After ~6 days, the maggots drop to the ground, burrow into the soil and pupate. Adults typically emerge 7 to 9 days later. There may be eight or more generations per year and each female can produce 500 to 600 eggs so the biotic potential and risk for myiasis is substantial. See also 'Myiasis treatment and prevention'.

Maggot of *Chrysomya rufifacies*. Source Austinh37/Wikipedia. CC BY-SA 3.0. Available from https://en.wikipedia.org/wiki/Chrysomya#/media/File:Chrysomya_rufifacies_Larva_1.JPG

Screwworm flies
(Family Calliphoridae)

There are both new and old screwworm flies that are superficially similar in appearance and have similar behaviours. Larvae of both groups require living tissue in which to develop, and ungulates, including cattle, are their primary hosts. Female screwworm flies seek out open wounds on the host to lay six to eight batches of creamy white eggs. The female feeds on liquids exuding from the wound, then lays the first eggs in shingle-like fashion along the border of the wound. Subsequent batches of eggs are laid every 3 to 4 days, and each female produces up to 400 eggs. The eggs hatch 8 to 24 hours later and immediately begin feeding on fluids in the wound. The larvae then burrow into and eat the underlying tissue causing extensive damage.

Screwworm larva (*Cochliomyia hominivorax*). Source: Heather Stockdale Walden, University of Florida

Larvae pass through three stages (or instars) over 5 to 7 days. Early instars are difficult to see and their movements are difficult to detect. However, by the third day, the larvae are readily visible, and it is possible to see up to 200 tightly packed, vertically oriented larvae embedded deep in the wound. When disturbed they tend to burrow deeper into the wound. Infected wounds often exude a discharge and have a distinct, unpleasant odour. Sometimes the larvae form an extensive pocket beneath a small wound opening. The terminal, third instar larvae exit the wound, drop to the ground, burrow into the soil and pupate. Pupation takes ~7 days and adults are

Screwworm fly (*Cochliomyia hominivorax*). Source: Judy Gallagher/Flickr. CC BY 2.0. Available from https://www.flickr.com/photos/52450054@N04/33364645563/

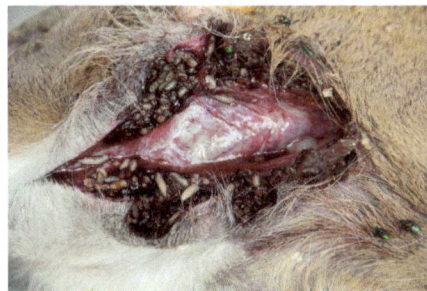

Open wound from screwworm. Source: Dr Holland, United States Navy.

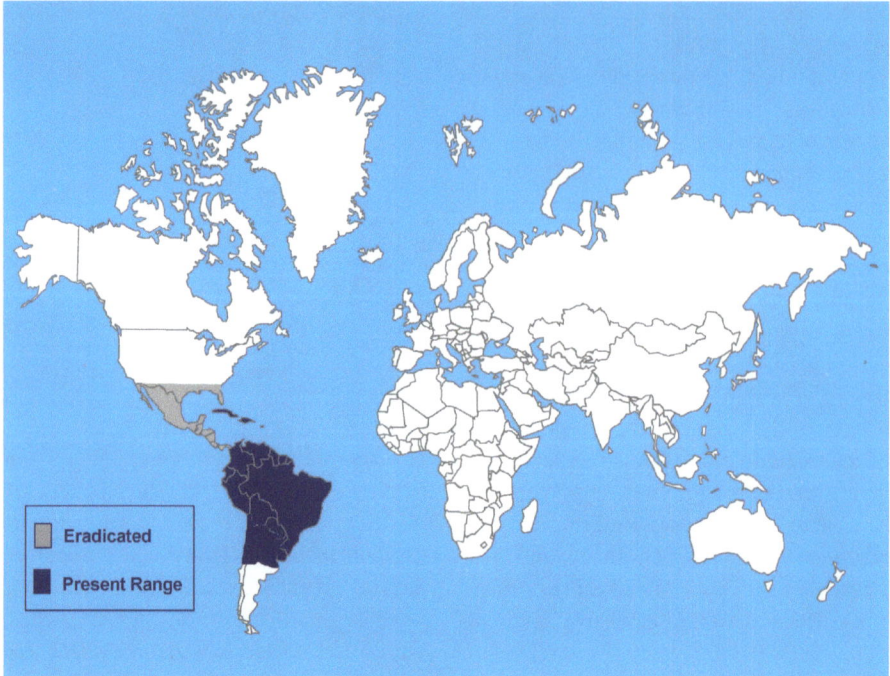

Distribution of screwworm fly (*Cochliomyia hominivorax*).

ready to mate in 3 to 5 days after they emerge from their pupa. Females only mate once, live on average 10 days, and can fly 10–20 km (6–12 miles) in warm, humid environments and up to 300 km (186 miles) in arid environments.

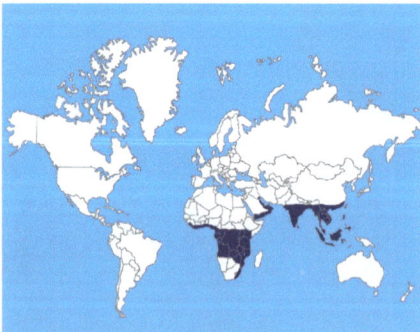

Distribution of Old World screwworm fly (*Chyrsomya bezziana*).

The Old World screwworm fly (*Chrysomya bezziana*) is tropical and the most important myiasis producing fly throughout Africa, the Middle East, Asia, some Indo-Pacific islands and India (see 'Sheep maggots'). This species ranks second only to tsetse flies as a pest of cattle in central and southern Africa. This species attacks wounds on various body parts, but the eyes, nasal cavities and head wounds are most frequently infested.

The New World screwworm fly (*Cochliomyia hominivorax*) is endemic to the Western Hemisphere where it has historically produced devastating economic losses due to damaging and/or killing livestock. Historically, this species was distributed from the southern United States southward throughout South America, but eradication efforts now limit its range from about Panama through South America, excluding Chile. Populations also exist in Cuba, the Dominican Republic, Haiti, Jamaica, and Trinidad and Tobago. See also 'Myiasis treatment and prevention'.

Spotted flesh fly (Family Sarcophagidae)

Human myiasis caused by the spotted flesh fly (*Wohlfahrtia magnifica*) is relatively rare. This species is broadly distributed in Europe, North Africa, the Middle East and Asia. Spotted flesh fly larvae require mammal tissue to survive. Human flesh is suitable, but they primarily infest livestock, mainly sheep, but also goats, horses, camels and cattle. The diurnal flies are most active from May to October. The dark coloured female fly is ~ 10 mm (0.39 inch) long, feeds on flowers and is larviparous, depositing 80 to 120 parasitic larvae onto hosts during her life. In humans, they deposit larvae in open wounds and orifices (i.e. mouth, nose, ears and eyes). The small larvae immediately burrow deeply into the tissue and are not easily detected with a simple visual exam. The larvae feed for 4 to 8 days, maturing through three larval stages, exit their host, drop to the ground, burrow

into the soil and pupate. Adults emerge 4 to 12 days later or they overwinter as pupae to emerge the next spring. See also 'Myiasis treatment and prevention'.

Other notable myiasis flies

Humpbacked flies (Family Phoridae), especially *Megaselia scalaris*, are tiny, dark-coloured flies that are occasionally implicated in passive human myiasis. Their preferred breeding habits are often hidden, difficult to find, decomposing plant or animal debris including liquefied garbage, sewage and carrion. They are attracted to areas of the body that produce odour or to wounds exhibiting infection or decay, which they infest. Similarly, the lesser house fly (*Fannia canicularis*) and the latrine fly (*F. scalaris*) (Family Fanniidae) are commonly found worldwide, and they occasionally cause intestinal and urinary tract myiasis. These flies normally prefer excrement and decaying organic material, and they commonly breed in latrines and cesspools. Such infestations are self-limiting, but they may produce psychological distress in the patient. The green bottle fly (*Lucilia sericata*) and the sheep blow fly (*L. cuprina*) (Family

Flesh fly (*Sarcophaga* sp.). Source: Frank Vassen/ Flickr. CC BY 2.0. Available from https://www.flickr. com/photos/42244964@N03/35499330022/

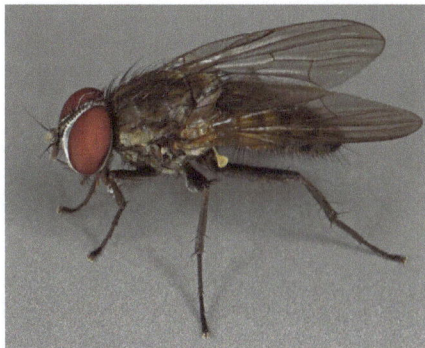

Lesser house fly (*Fannia cannicularis*). Source: Janet Graham/Flickr. CC BY 2.0. Available from https://www. flickr.com/photos/130093583@N04/16737966188/

Green bottle fly (*Lucilia serricata*) on a chipmunk carcass. Source: David E. Bowles

Calliphoridae) feed on decaying organic matter including faeces, dead animals and infected wounds of living animals. Both species are distributed worldwide, particularly in warmer areas. Cases of human myiasis attributed to these species typically occur when the flies invade unprotected wounds. Maggots of *L. sericata* have

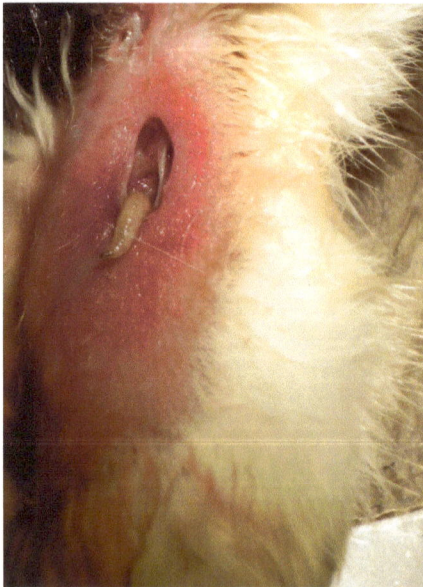

A fly maggot in the wound of a cat. Source: Uwe Gille/Wikipedia. CC By 3.0. Available from https://en.wikipedia.org/wiki/Myiasis#/media/File:Myiasis-cat.jpg

been successfully used in maggot therapy where sterile larvae are intentionally introduced into wounds to help manage infected or debride necrotic tissue. Such therapy is reported to be pain free. See also 'Myiasis treatment and prevention'.

Myiasis treatment and prevention

Understanding area risk for contracting myiasis and proper use of personal protection measures (see 'Personal protection measures') are crucial for avoiding it. In general, avoidance, barriers and good sanitation measures are key to preventing accidental and facultative myiasis. Travel and/or adventures to high-risk areas/habitats, particularly those involving extensive time outdoors, around herd animals, around burrow-dwelling animals or in indigenous dwellings (dirt floors, mud huts, thatched or woven-stick construction, etc.) place people at highest risk for infestation. If sleeping outdoors, tents with window screens, bed nets treated with permethrin or a permethrin treated sleeping bags are especially helpful. Practise good field site waste disposal, fly prevention and control to reduce fly populations and infestation risk. Only well-cooked or factory sealed foods should be eaten unless food safety is absolutely certain.

Open wounds should be immediately and thoroughly cleaned, sanitised and dressed (bandaged) to prevent access by flies. Even with those precautions, wounds should be monitored daily for symptoms of myiasis (larvae, movement, wound expansion, etc.). Most cases of myiasis are uneventful, self-resolving and rarely fatal, but medical intervention is recommended if symptoms of myiasis are observed. Medical assistance should always be sought if the myiasis involves the eyes. A common method for getting maggots to leave a wound is to eliminate their air supply by applying copious amounts of petroleum jelly or similar substance over the wound, which forces them to the surface where they can be physically removed. Sometimes a practitioner may use local anaesthesia and extract the maggots surgically. Regardless of the removal method, follow-up examination is typically required to ensure all of the maggots were removed and that secondary infection did not occur.

Biting flies and gnats

Biting flies require a blood meal to produce viable eggs. Their mouthparts are modified to pierce or lacerate tissue, and many allow a pool of blood to form, which they then consume. Most mosquito species insert their mouthpart stylets into a host's small blood vessels and feed by drawing blood from them, aided by an organ called a cybarial pump. In addition to being painful and annoying, such bites can also lead to allergic reactions and result in transmission of potentially fatal disease agents. We cover the major groups of biting flies, how to prevent bites, and their treatment.

Mosquitoes
(Family Culicidae)

Mosquitoes are the most serious invertebrate threat worldwide. Besides causing annoying, sometimes painful, bites, they transmit dangerous and often fatal human and animal pathogens that cause diseases such as malaria, filariasis, yellow fever, dengue, chikungunya, zika and various encephalitides (Appendix 7). With ~3500 species in 41 genera worldwide, we do not attempt a comprehensive review here. However, understanding some key biological characteristics can help you recognise potential mosquito threats (Appendices 6, 7) so appropriate protection measures can be taken. Threat reduction and personal protection from mosquitoes is crucial in some areas where mosquito-borne diseases are potentially life threatening. Therefore, we briefly present that information, and then cover medically important genera, specific species, habitats, biology, and associated diseases.

Mosquito biology. The complete mosquito life cycle (egg, larva, pupa and adult) takes ≤14 days, but environmental conditions (temperature, humidity, moisture, etc.) and species-specific characteristics affect the longevity of each life stage. Additionally, the various mosquito genera have highly variable biologies, including egg laying methods and modes, feeding and activity patterns, and resting locations. For example, mosquitoes in the genera *Culex* and *Culiseta* species lay up to 200 eggs in floating rafts. By comparison, other genera (*Aedes, Anopheles, Ochlerotatus* and *Psorophora*) lay eggs singly. *Aedes, Ochlerotatus* and *Psorophora* lay eggs on damp soil that will flood and many use small artificial and natural containers (e.g. old tyres, bottles, cans, birdbaths, tree holes) attaching eggs to surfaces (e.g. walls, stems) just above the water line.

Floating egg raft of *Culex* sp. Source: Harold J. Harlan

Aedes vigilax larvae and pupa. Source: Stephen L. Doggett

Culex annulirostris larvae. Source: Stephen L. Doggett

Most mosquito eggs hatch within 48 hours after being laid. The larvae, commonly called wigglers, go through four stages (instars) over ~7 days (depending on temperature), becoming progressively larger with each moult. Larvae are exclusively aquatic and usually prefer standing fresh or brackish water. However, *Anopheles* can inhabit backwaters of streams. Most larvae have a respiratory siphon on their abdominal tip with a spiracle located at its tip. The larvae obtain atmospheric oxygen through this siphon, which is accomplished by inserting the siphon through the water surface and hanging upside down, oriented vertical to the surface. In contrast, *Anopheles* larvae do not have a prominent siphon, and they must lie parallel to the water surface to get oxygen through an elevated spiracle. The genera *Coquillettidia* and *Mansonia* larvae have modified respiratory siphons that they use to pierce aquatic plant tissues from which they get air. Most mosquito larvae feed on microorganisms and organic matter, but some genera, such as *Toxorhynchites*, are predacious and feed on other insects, including mosquitoes. The pupal stage, commonly called a tumbler, is a mobile, non-feeding, developmental stage that actively tumbles when disturbed.

Emerging adult mosquitoes must rest on the water surface a short time to get dry, expand their wings, and harden their exoskeleton before they can fly. During this time the newly emerged mosquito is vulnerable to wind and water movement, predators and other threats. They require a

Anopheles annulipes larva. Source: Stephen L. Doggett

Coquillettidia xanthogaster larvae. Source: Stephen L. Doggett

Emerging adult mosquito with larval mosquitos in surrounding water. Source: NY State IPM Program/ Flickr. CC BY 2.0. Available from https://www.flickr.com/photos/99758165@N06/14748337176/

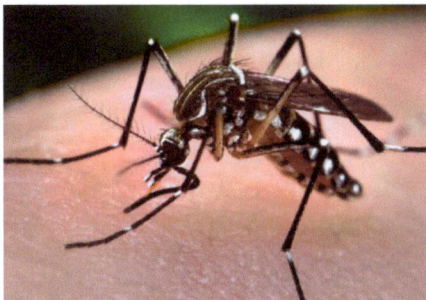

Aedes aegypti. Source: James Gathany, US Centers for Disease Control and Prevention

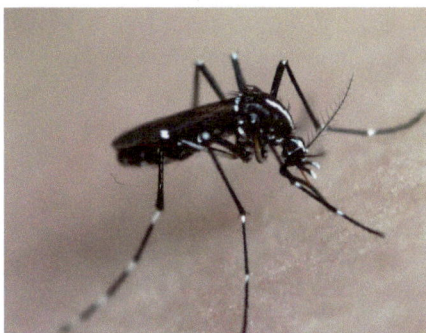

Aedes albopictus. Source: Frankieleon/Flickr. CC BY 2.0. Available from https://www.flickr.com/photos/armydre2008/3779182508/

couple of days after emergence before they are ready to mate. Both males and females feed on plant nectar or other suitable sugar sources, and only female mosquitoes require a blood meal that serves as protein source to produce eggs. Female mosquitoes feed on a wide range of different animals including mammals, birds, reptiles and amphibians. A variety of external stimuli, or their combinations, serve to influence female biting behaviour and attract them to the host, including carbon dioxide, temperature, moisture, smell, colour, movement and other chemical cues. One exception is *Toxorhynchites* where both sexes only feed on nectar. The average lifespan for most female mosquitos is around 2 weeks, but some may live for up to 5 months in favourable circumstances.

Genera *Aedes* and *Ochlerotatus*. Mosquitoes in these closely related genera are painful, persistent biters. There is a prevalence among most of the species to feed during morning hours or at dusk (called crepuscular feeding). Other species are diurnal (daytime) feeders, preferring cloudy days and shady areas, or at dawn and dusk. *Ochlerotatus triseriatus* is the primary vector of La Crosse encephalitis in North America. *Aedes*, particularly *Ae. aegypti* and *Ae. albopictus*, in addition to being major nuisance bitters, are important vectors of mosquito-borne diseases in many areas of the World. They transmit several dangerous viral diseases (e.g.

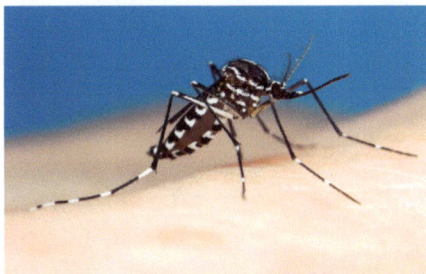

Aedes scutellaris. Source: Stephen L. Doggett

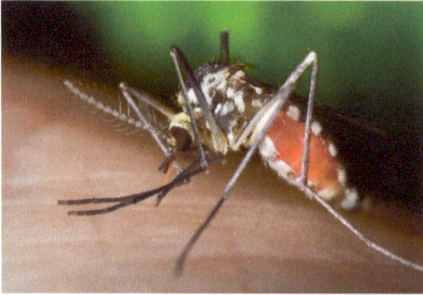

Ochlerotatus triseriatus. Source: Animal Diversity Web/Flickr. CC BY 2.0. Available from https://www.flickr.com/photos/adwsocial/14919624442/

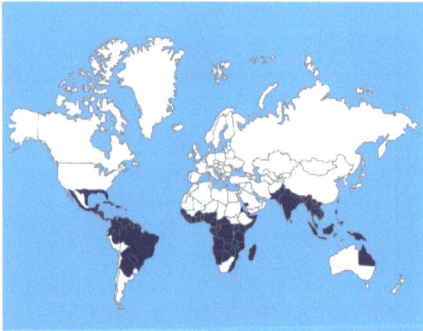

Distribution of Yellow fever mosquito (*Aedes aegypti*).

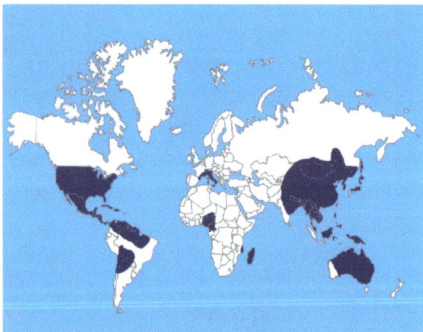

Distribution of Asian tiger mosquito (*Aedes albopictus*).

dengue fever, yellow fever, chikungunya, Zika) in tropical and subtropical areas worldwide. Many species of *Aedes* require very little water in which to lay their eggs. They preferentially select containers such as tree holes and other natural cavities, and fabricated objects such as tyres, bottles, and flowerpots). Accordingly, members of this genus are often found around homes and other structures where favourable habitat can be found, and they readily enter human dwellings to rest and feed.

Genus *Anopheles*. *Anopheles* mosquitoes are the vectors of human malaria, making them among the most important disease vectors in the world. Of the ~430 known species of *Anopheles* worldwide, only 30 to 40 of them transmit malaria. *Anopheles* larvae inhabit fresh- or saltwater marshes, mangrove swamps, rice fields, grassy ditches, stream or river edges, and small temporary rain pools. Many prefer vegetated aquatic habitats while others prefer open water with no vegetation. Some breed in open, sun-lit pools while others prefer shaded forest sites. *Anopheles* females are easily distinguished from those of other genera because they point their abdomen nearly directly upward while resting, rather than parallel to the surface. Additionally, the larvae do not have a long respiratory siphon, and they lie parallel to the water surface so their posterior spiracle can make

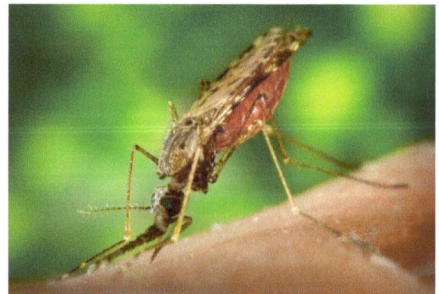

Anopheles albimanus. Source: James Gathany, US Centers for Disease Control and Prevention

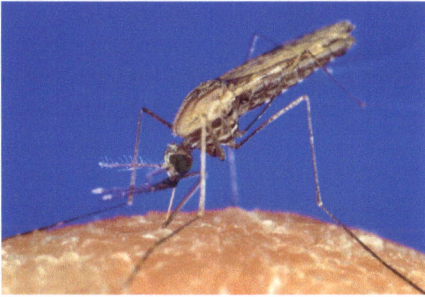

Anopheles gambiae. Source: James Gathany, US Centers for Disease Control and Prevention

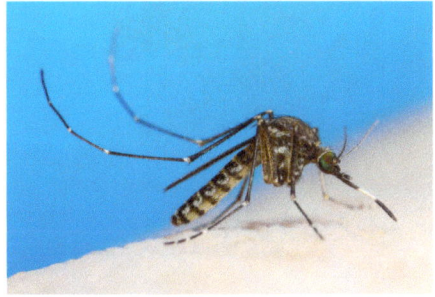

Culex annulirostris. Source: Stephen L. Doggett

Distribution of *Anopheles* vectors of malaria.

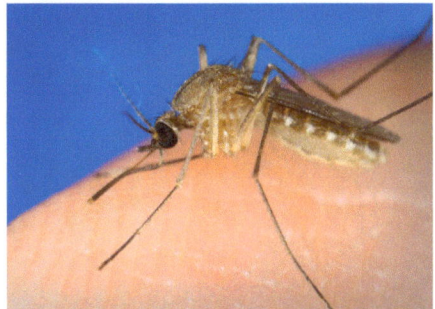

Culex quinquefasciatus. Source: James Gathany, US Centers for Disease Control and Prevention

contact with the atmosphere. Malaria is the most important vector-borne disease in the world. The WHO International Association for Medical Assistance to Travelers (IAMAT) site (https://www.iamat.org) maintains a current 'World Malaria Risk Chart' that covers country-by-country malaria risks, chemoprophylaxis, *Anopheles* vectors, their biology and habits.

Genus *Culex*. These mosquitoes are persistent biters, and they primarily feed at dusk or at night. They are distributed worldwide. Domestic and wild birds are their preferred hosts, but they will feed on various mammals, including humans. Because they tend to feed on a variety of different host types, many species can transmit several dangerous viruses

(Appendix 7). Females lay eggs in organically enriched water, but they use various aquatic habitats. Most *Culex* species are relatively weak fliers that generally remain close to their breeding sources. The adults will readily enter human dwellings to rest and feed. Adults may live a few weeks during warm summer months, but females emerging in late summer may seek sheltered areas to overwinter (hibernate) until spring. Some species even breed and overwinter in the sewers of large cities.

Other mosquitoes. Other mosquito genera are primarily annoyance biters and/or only secondary vectors of human disease. *Culiseta* mosquitoes are moderately aggressive mosquitoes that can present a substantial annoyance problem

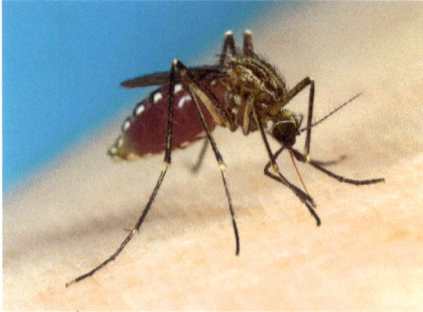

Coquillettidia linealis. Source: Stephen L. Doggett

Psorophora ferox. Source: Katja Schultz/Wikipedia. CC BY 2.0. Available from https://en.wikipedia.org/wiki/Psorophora#/media/File:Psorophora_ferox.jpg

Biting midges (Family Ceratopogonidae)

Biting midges, also known as punkies or no-see-ums, are annoying, painful biters that are distributed worldwide. Even with small population densities, these insects can easily interrupt outdoor plans if personal protection measures are not used. Most biting midges are pests of other animals, including insects, but females of two genera *Culicoides* (~1000 species) and *Leptoconops* (~90 species) readily take human blood meals. In addition, certain *Culicoides* appear to transmit Oropouche (a febrile virus) in Brazil as well as three non-pathogenic filarial parasites in Africa, the Caribbean, and Central and South America (see Appendix 7). Certain species do transmit important parasites and/or viruses to horses and sheep. Biting midges do not fly vary far, so most bites occur near their various aquatic and semi-aquatic breeding sources including tree holes, decaying vegetation, mud banks of streams, lakes

when their populations are large. They feed primarily during evening or the daytime in heavily shaded areas. Representatives of the genera *Haemogogus* and *Sabethes* have been implicated in the transmission of yellow fever in rural Central and South America and tropical Africa, but they are not pests associated with large human populations. The genera *Coquillettidia* and *Mansonia* are occasionally mainly nuisance biting mosquitoes, and they are becoming more pestiferous due to expansion of human populations into their natural habitats. Floodwater mosquitoes in the genus *Psorophora* can be major pests in areas that support large populations, such as rice-growing areas.

Biting midge (*Culicoides* sp.). Source: Stephen L. Doggett

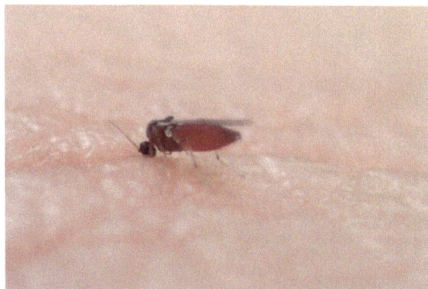

Biting midge (*Culicoides* sp.) taking a blood meal. Source: D. Sikes/Flickr. CC BY 2.0. Available from https://www.flickr.com/photos/alaskaent/8036621825/

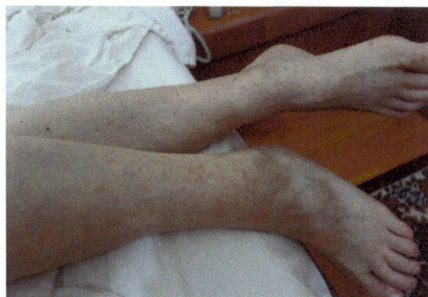

Bite marks from biting midges often termed 'sand flea bites'. Source: Larnie & Bodil Fox/Flickr. CC BY 2.0. Available from https://www.flickr.com/photos/larnie/8592115242/

and wetlands, tidal flats and salt marshes. *Leptoconops* is typically diurnal while *Culicoides* can be either diurnal or nocturnal. Diurnal species mainly feed in early morning and late afternoon, rather than at the hotter times of day. In addition to the painful bites, salivary components transmitted during feeding can cause localised dermal reactions that may lead to dermatitis, or secondary infection if they are scratched. These dermal reactions may appear as bumps or nodules on the skin that itch intensely, and they may take months to disappear. These small (2–3 mm, 0.08–0.12 inch) flies can pass through window screens and standard mosquito netting unnoticed. Therefore, personal protection measures, especially repellents, are essential for preventing bites from these midges.

Sand flies
(Family Psychodidae, Subfamily Phlebotominae)

Similar to biting midges, sand flies are tiny (<2 mm or <0.08 inch). Because of their small size, they often go unnoticed until they bite. Sand flies are painful, annoying biters. They are poor flyers and commonly hop on solid substrates. Their characteristic 'hairy' appearance is due to their heavily setose body, and, when resting, they hold their wings in a V shape above their abdomen. In addition to annoying bites, sand flies can transmit three serious diseases: sand fly fever, bartonellosis, and leishmaniasis (see Appendix 7). Similar to mosquitoes, male sand flies do not bite, and females require one or more blood meals per egg batch (15–80). Females feed at night on various warm- and cold-

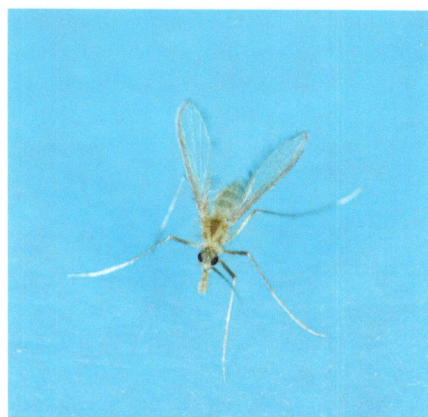

A sand fly (*Phlebotomus papatasi*). Source: US Department of Agriculture/Flickr. CC BY 2.0. Available from https://www.flickr.com/photos/usdagov/8412913936/

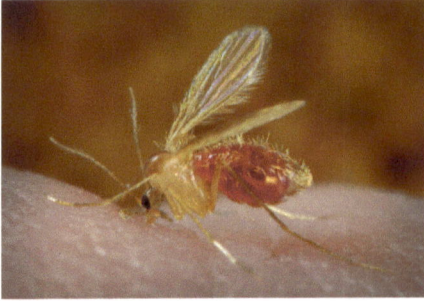

A sand fly (*Phlebotomus papatasi*) taking a blood meal. Source: James Gathany, US Centers for Disease Control and Prevention

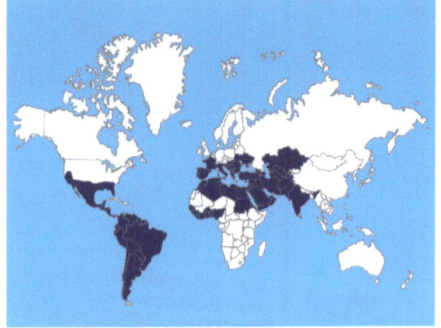

Distribution of sand flies (Phlebotominae).

blooded animals including cats, dogs, rodents, birds, bats, herd animals, reptiles and people. There are four larval instars, which require high humidity in habitats such as tropical forests, rubble, caves, rock piles and animal burrows where they feed on organic particles. Drier conditions are required for pupation. Sand flies are widely distributed including the Americas, Africa, around the Mediterranean, India, Near East, Middle East and China (see Appendix 6). Occasionally large populations emerge, which can make outdoor activities nearly impossible. The two most medically important genera are *Lutzomyia* (New World) and *Phlebotomus* (Old World). In

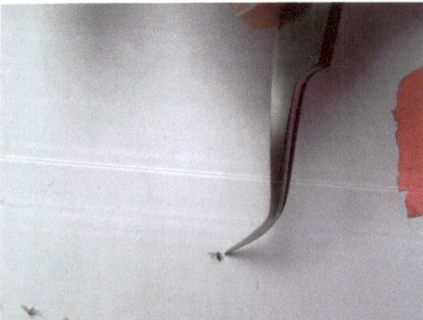

Insert Adult sand fly showing their small body size. Source: Tod Vinto, United States Army

Australia, New Zealand and the United States, the general name 'sand fly' sometimes is applied to any small biting fly inhabiting coastal sandy areas and mangroves, but these insects are not true sand flies and most often they belong to the Family Ceratopogonidae (see earlier).

Black flies
(Family Simuliidae)

There are over 1800 black fly species among over 30 genera, and they occur in temperate and boreal regions worldwide. Most of the known species belong to the genus *Simulium*. Black flies can be significant and medically important human pests. Adults are small, 3–4 mm (0.12–0.16 inch), dark-coloured, robust, hump-backed flies, thus the name buffalo gnats. Other common names used for these flies have included turkey gnat, Golubatz fly, moshki, borrachudos, cseszlék, botlass and kurukundu.

Male black flies feed on nectar, while females require blood to nourish their eggs, which they do during daytime feeding, primarily at dawn and dusk. The female black fly penetrates the host's skin with knife-like mouthparts and ruptures fine capillaries. They then imbibe the

Black fly larvae (*Simulium* sp.). Source: Esteban Peláez Sánchez, Laboratorio Limnología UCO/Flickr. CC BY 2.0. Available from https://www.flickr.com/photos/129724881@N03/16253393930/

Black fly larvae aggregated in a stream. Source: Ask.extension.org. CC BY-SA 3.0 US. Available from https://ask.extension.org/uploads/question/images/attachments/000/085/732/20151214_152159_original.jpg

pooled blood. Black flies feed on a broad range of birds and mammals, including people. Although bites are generally painful, salivary anticoagulants produced can have a numbing effect in some cases. In addition to being painful, bites cause localised swelling (oedema), inflammation and intense itching, which can last several days. Severe attacks can cause swelling of the limbs or 'black fly fever,' which is characterised by headache, nausea, fever, swollen lymph nodes and aching joints. Additionally, some people may suffer severe allergic reactions requiring hospitalisation. In addition, some species transmit onchocerciasis, or river blindness, which is the second leading cause of blindness after trachoma in some areas. Most cases of onchocerciasis occur in equatorial, sub-Saharan Africa. Fewer cases occur in Yemen, and isolated areas extending from southern Mexico, southward through mountainous regions of northern South America. Although a serious disease, visibly detectable onchocerciasis (blindness) is obtained only following many black fly bites.

Once fed, most female black flies lay a single egg batch of 200 to 500 eggs in fresh, clean, flowing water or wet surfaces of aquatic vegetation. The characteristically shaped larvae and pupae of most black flies inhabit oxygen rich, non-polluted, rapidly flowing water sources (e.g. large rivers, icy mountain streams, trickling creeks, waterfalls). In contrast, the immature stages of *Cnephia* spp. inhabit slow-flowing streams and swamps.

Eggs hatch in 4 to 30 days, or rarely longer. Larvae may have four to nine larval instars that take 1 to 6 months to develop, and they grow to 5 to 15 mm (0.13–0.40 inch) long. Larvae consume organic matter or small invertebrates carried to them on stream currents, which they catch with a pair of elaborate fan-shaped structures on their head. The pupal stage lasts 4 to 7 days, and they may or may not be located within a silken 'cocoon.' The emerging adults live several weeks and are strong flyers, dispersing, unaided, considerable distances (tens to hundreds of kilometres/miles). Non-biting males and females can form massive swarms making outdoor activities unpleasant or intolerable. However, they normally do not enter human structures. Most major pest species complete several generations per year, but

Adult black fly (*Simulium* sp.). Source: Stephen L. Doggett

some only have one generation, especially in cold climates.

Repellents offer some protection against black flies, but the best personal protection is to wear a head net, long sleeves and pants, especially where large swarms occur.

Tsetse flies
(Family Glossinidae)

Tsetse flies (Tswana meaning 'fly'; also tsetse or tik-tik flies), genus *Glossina*, are yellow-, grey- or dark brown, large (8–17 mm, 0.2–0.6 inches) flies. Tsetse flies are readily recognisable because they fold their wings over their backs, and their long proboscis projects forward approximately one-half their body length beyond their head. About 20 species and seven subspecies have been described from throughout central Africa between the Sahara and Kalahari deserts (14°N–29°S).

Tsetse flies are often arranged in three groups including: (1) 'forest tsetse' (*Glossina fusca* complex) living among the region's forests and who rarely bite people; (2) 'savanna tsetse' (*G. morsitans* complex) occurring in east and central African thornbush; and (3) 'riverine tsetse' (*G. palpalis* complex), which inhabit the edges

of wet woodlands and central African rivers.

Tsetse flies have an interesting life cycle. Female tsetse flies live ~14 weeks (males 6 weeks). They mate only once, but store the sperm and fertilise each egg individually. They carry the larva internally in a uterus. First instar larvae are hatched about every 10 days throughout the life of the female, and several generations may be produced in a year. The first egg is fertilised 7 to 9 days after the female emerges from the pupal stage, and the first larva hatches ~3 to 4 days later. The larva matures over 8 to 12 days through three instars while feeding on a specialised milk gland secretion produced by the female. The female seeks sheltered, sandy soil to deliver a fully developed larva. The larva does not feed, but it burrows under the soil to pupate, transforming over 3 to 5 weeks (8 to 10 weeks during cooler weather) into an adult. Because the females must provide all the nutrients to complete the entire life cycle, they must feed every 2 to 5 days. Both males and females require blood, and they use visual cues to locate their hosts. Feeding occurs during the daytime causing painful bites and irritating, itchy wounds. All tsetse flies feed primarily on warm-blooded mammals, including people, and occasionally large reptiles (crocodiles).

Twenty-three species of *Glossina* transmit African sleeping sickness (Rhodesian and Gambian African trypanosomiasis). They acquire a life-long infection while feeding on infected hosts and transmit the pathogen mechanically by regurgitating infected blood from a previous meal, or biologically through development of the parasite (*Trypanosoma brucei*) in the tsetse fly, which is transmitted during feeding. Rhodesian sleeping sickness is transmitted by tsetse flies in the *G. morsitans* complex, and it is distributed in throughout the eastern/southern African savanna (i.e. Malawi, Tanzania, Uganda and Zambia) affecting both humans and animals. In contrast, Gambian sleeping sickness is transmitted by tsetse flies in the *G. palpalis* complex, which is distributed along river systems throughout western and central Africa (i.e. Angola, Central African Republic, Chad, Democratic Republic of Congo, Republic of Congo, Sudan and Uganda). Gambian sleeping sickness is restricted to humans.

The risk of contracting sleeping sickness in Africa is generally low, but the greatest risk occurs while visiting rural areas, living outdoors, around thick vegetation or touring game parks. Tsetse flies are highly attracted to moving vehicles and dust clouds, so vehicle-based safari tours increase the risk of exposure. Prolonged visits increase the chance of becoming infected. The initial symptoms of sleeping sickness are flu-like, including fatigue, high fever, headaches, joint pain, muscle aches and itching, which eventually progress to swollen lymph nodes. Such

Distribution of tsetse flies (*Glossina* spp.).

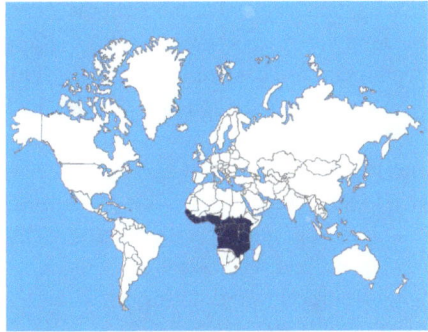

Tsetse fly (*Glossina* sp.). Source: Oregon State University/Flickr. CC BY-SA 2.0. Available from https://www.flickr.com/photos/oregonstateuniversity/10040375154/

symptoms may develop within 3 days or be delayed weeks, months or even years, depending on the pathogen strain. The pathogen eventually infects the nervous system causing severe symptoms, including disrupted sleep patterns, mood swings, reduced coordination and confusion. Left untreated, it can be fatal.

Stable flies
(Family Muscidae)

Unlike most Muscidae, the genus *Stomoxys* feeds on mammalian blood. *Stomoxys* contains 18 known species, but *Stomoxys calcitrans* (dog fly, barn fly, biting house fly), *S. nigra, and S. sitiens* are the primary human pests. Originally European, stable flies now occupy temperate and tropical areas worldwide, while *S. sitiens* has an African and oriental distribution, and *S. nigra* is African. Among these, the stable fly, *S. calcitrans,* can be a formidable pest.

The stable fly primarily attacks a wide range of ungulate hosts (e.g. cattle, horses, mules, camels, goats, sheep), but also swine, cats, dogs and people. The brownish grey stable fly has a greenish yellow sheen and resembles a house fly. However, unlike the house fly, it has four dark longitudinal thoracic stripes, dark abdominal spots forming a chequerboard pattern and a prominent, black, piercing proboscis. Both males and females require blood. Their breeding habitats are moist, animal dung enriched, fermenting organic materials (e.g. straw, hay, waste silage, animal bedding, decaying vegetable or fruit matter, animal feed, seaweed, grass clippings, compost heaps). Biting attacks occur mostly outdoors near domestic animals or along beaches, during daylight hours (normally morning and late afternoon), and they target your elbows, legs and knees. Their bites easily penetrate thin fabric, socks and stockings. During high population periods or rainy weather, they may enter human dwellings. Females take 2 to 5 minutes to get enough blood (three times their bodyweight) to develop eggs. However, they are easily disturbed and usually feed several times to get a sufficient blood meal. This behaviour makes them aggressive, persistent biters. They can inflict significant pain and irritation, especially when present in high densities. Furthermore, although extremely rare, they may also mechanically transmit diseases (leishmaniasis, trypanosomiasis and certain bacteria) via contaminated mouthparts or feet.

The entire life cycle of the stable fly, egg to adult, is highly variable depending on temperature, and it may take from 12 to 36 days to complete. After feeding, females seek a suitable breeding habitat, lay four to five clutches (23–100 eggs each), then feed again, repeating this sequence 10 to 11 times to lay between 200 to 800 eggs. The eggs hatch within 1 to 5 days. The larvae (maggots) grow through three instars over 6 to 26 days, and then seek a drier environ-

Stable fly (*Stomoxys calcitrans*). Source: Eran Finkle/Flickr. CC BY 2.0. Available from https://www.flickr.com/photos/finklez/3734405370/

ment to pupate. The pupae develop over 5 to 26 days before emerging to become adults. Emerging males feed and die soon after mating, but females begin the egg laying cycle 5 to 10 days later and may live 20 to 40 days. Multiple generations may be produced in warmer climates. Adults are strong fliers, and they are ready to fly within an hour after emergence. They can travel long distances (5–225 km; 3–140 miles) in search of new habitats and food sources. Stable flies overwinter in cold climates as larvae and pupae.

Horse flies and deer flies (Family Tabanidae)

The Family Tabanidae contains an estimated 4300 species in 137 genera. Horse and deer flies are closely related, have a similar appearance and share similar biologies, but deer flies are smaller than horse flies. All tabanids are relatively large flies up to 32 mm (1.25 inches) long, and their colouration is usually drab brown, black, grey or yellowish. The large, transparent, fan-shaped wings are mottled or have dark patches or stripes. Tabanids have broad, robust bodies and distinctive, bulging compound eyes. Many tabanids have brightly coloured, iridescent eyes that are sometimes spotted or striped while they are alive.

Tabanid fly life cycles vary considerably depending on the species. They live 70 days to 3 years (usually 1 year), but most of their life is spent in the larval stage and they often overwinter in this life stage. Adults typically emerge in late spring through summer and live 27 to 60 days. Male and female tabanids feed on plant nectar, plant sap, or aphid and scale insect honeydew, but females of most species require a blood meal for proper egg development. Some tabanids actually mate while flying, while others land first. Females may attack several hosts to get enough blood and then lay a single egg mass on the underside of leaves, tree branches, logs or rocks over-hanging water, or on wet soil or mud. The eggs hatch ~2 to 7 days later and the larvae drop to the ground or water surface to develop. Depending on the species, tabanid larvae may inhabit streams and ponds, damp forest soils, moist decomposing wood and fresh- or saltwater marshes, while deer fly larvae are largely aquatic, preferring marshes, ponds and streams. Larvae are predatory on a variety of small organisms.

Tabanids feed on wildlife, livestock and humans throughout tropical and temperate environments worldwide, except Hawaii, Greenland and Iceland. They bite exclusively during daylight hours, and most commonly at dawn and dusk or in shaded areas. Tabanids are first attracted to a potential host by respiratory gases and other compounds produced by the body (e.g. carbon dioxide, octenol), and they secondarily use visual cues (movements, size, shape, colour) to focus their attack. They are especially attracted to dark colours (dark blues, reds, browns, blacks) and shiny surfaces. This combination of attractants also makes slowly moving automobiles attractive to them and they will readily fly in open windows. Once they locate a potential host, they bite any exposed skin, particularly around the ankles, arms, head and neck. They use large, biting-slashing mouthparts to lacerate the skin into which they inject an anticoagulant, creating deep, painful, bleeding bites. The subsequent irritation and swelling may last 1 to 2 days. They are aggressive, tenacious biters that quickly return if their feeding attempts are interrupted. Tabanids are strong, fast, long distance fliers and actively pursue potential hosts. However, they normally do not enter structures. Tabanids are most abundant and commonly encountered on hot summer days near lake or pond edges, stream banks, wetlands, moist forests, seepage areas, swamps, moist debris or other bodies of water.

The most commonly implicated human biting horsefly pests belong to three genera and include the horseflies *Tabanus* and

Deer fly (*Chrysops* spp.). Source: Gary Alpert, US Centers for Disease Control and Prevention

Hybomitra, and the deer fly genus *Chrysops*. All three genera contain species that are aggressive feeders. Tabanid bites are painful and annoying, and they also present a modest risk for disease transmission in some areas. African deer flies are mechanical vectors of parasitic filarial worms (i.e. *Loa loa*). The two most important *Chrysops* species involved in transmitting *Loa* are *C. silacea* of equatorial African, and *C. dimidiata* of west and central Africa (particularly Nigeria and Cameroon). Similarly, the North American deer flies *C. discalis*, *C. fulvaster* and *C. aestuans* have been implicated as vectors of tularemia (deer fly fever, rabbit fever), which causes skin ulcers, swollen lymph nodes, fever, pneumonia and headache. North American horse flies also may transmit anthrax. In addition to transmitting pathogens, tabanid salivary juices can stimulate allergic reactions (hives, wheezing, etc.) and the open, bleeding bites are susceptible to secondary infections.

Repellents are not particularly effective against tabanids, so situational awareness and avoidance are key to preventing bites. Permethrin-treated protective clothing including, hats, head-nets, even neckerchiefs may discourage landing and feeding behav-

Horse fly (*Tabanus atratus*). Source: Mike Keeling/ Flickr. CC BY-ND 2.0. Available from https://www. flickr.com/photos/pachytime/2813943742/ x2kS2Q

iour. Wearing hats, long sleeves and pants greatly reduces bites from tabanids. See also 'Biting fly prevention and treatment'.

Biting fly prevention and treatment

Situational awareness, avoidance and personal protection are key to protection from biting flies and the diseases they may transmit. Travellers should assess the risk from biting flies that occur in the areas to which they may travel, to understand the relative risk of contracting vector-borne diseases (see Appendix 6 and 'Education' section). A physician should be consulted before travel to malarious areas so that appropriate chemoprophylaxis can be prescribed. Use guidance presented in Appendices 2 and 3 to select and properly use effective repellents (see Appendix 1 and 2), protective clothing, bed nets and other netting/screening. Bed nets and clothing should be treated with permethrin. When bites do occur, over-the-counter topical corticosteroids, systemic antihistamines and anti-pruritic (itching) ointments can reduce itching. To prevent secondary infections, try not to scratch bites.

Horse fly (*Tabanus stygius*). Source: Gary Alpert, US Centers for Disease Control and Prevention

Medical treatment should be sought at the earliest opportunity following onset of a febrile illness that develops during or after travel to high-risk areas. Upon return from travel to high-risk areas, we recommend seeking a follow-up medical evaluation because onset of several vector-borne diseases can develop well after exposure. Medical providers should be informed about international travel or other travel to high-risk areas so they can gauge any potential exposures to aid diagnosis. Depending on the specific circumstances, special laboratory tests, medications and/ or supportive care may be necessary.

Eye gnats (Family Chloropidae)

Gnats are not biting flies, but they are included here because their feeding behaviour may cause the sensation of being bitten. Some 2000 chloropid species in over 160 genera are described worldwide. They are small (<3 mm or 0.13 inch long), clear winged and variously coloured flies. The majority of chloropids are plant feeders, but a few are attracted to vertebrate animals. They are attracted to tears, sweat around eyes and natural body orifices (nostrils, mouth, anus and exposed genitals) or other body products (e.g. blood, pus, sebaceous material, faeces). Some species also frequent open wounds. When the flies are present in large numbers, they can be a major nuisance. Most of the nuisance species belong in the genera *Hippelates* (North America), *Liohippelates* (North through South America, Caribbean) and *Siphunculina* (mostly Asia, Afric, Australia, Europe). These genera also include some medically important species.

Eye gnat (Chloropidae). Source: D. Sikes/Flickr. CC BY-SA 2.0. Available from https://www.flickr.com/photos/alaskaent/273243744/

The complete life cycle for most species takes 11 to 90 days, depending on environmental factors such as temperature and moisture but typically lasts 21 to 28 days. Females require freshly disturbed, moist, organically rich (e.g. grass clippings, hay, manure), well-drained and aerated sandy soils. Digging, ploughing, harrowing or even livestock activities create such conditions. Newly hatched maggots burrow under the soil to feed on decomposing organic matter, developing through several stages (instars) over 7 to 11 days in warm weather. Pupation takes 6 to 10 days near the soil surface. Adults emerge and begin mating 5 to 8 days later among shrubs and brush.

Chloropids tend to be particularly abundant during autumn (fall) and spring. They can occur in large population densities under the right circumstances, and these extremely annoying and persistent pests quickly return when brushed aside. They are often drawn to shade or thick vegetation and become most active at dawn and dusk. Although they do not bite, the spines on their sponging/lapping mouthparts act like rasps that may cause minute scratches, particularly to the eye (ocular lesions). Larger species can actually produce a biting sensation.

Although primarily a physical nuisance, chloropids can serve as mechanical vectors of bacteria that cause infection in humans and other animals. For example, in the Caribbean and South America, chloropids have been implicated in the mechanical transmission of a spirochete (*Treponema pallidum pertenue*) that causes an ulcerative skin infection known as yaws. Some species of *Liohippelates* are known to transmit one or more bacteria that cause acute human conjunctivitis also known as 'pink eye' or 'sore eyes,' as well as streptococcal skin infections. In Brazil, these flies have been linked to acute bacterial (*Haemophilus influenzae*) conjunctivitis among children called Brazilian purpuric fever. This infection results in a sepsis that may be rapidly fatal.

Eye gnat prevention and treatment. Situational awareness, avoidance, repellents and head nets are excellent preventive measures against eye gnats. When possible, use finely screened (14–16 mesh) and/or sealed shelters to exclude eye gnats and keep doors, windows and vents of dwellings closed when practical. See 'Personal protection measures' section for additional details.

Filth flies
(Families Muscidae, Calliphoridae, and Sarcophagidae)

Filth flies breed in and feed on decomposing organic materials such as faeces, dead animals (carrion) and food wastes. They walk, feed, and defecate on these pathogen-contaminated materials. Most have sponging mouthparts and must consume liquids, which they regurgitate (vomit) along with digestive enzymes to further break down (liquefy) the food item, and then consume it once again. During this process, they also take up any pathogens (e.g. bacteria, viruses, parasites) that may be present, which also stick to their mouthparts, feet, legs and body setae. They also defecate contaminated faeces (fly spots) while feeding. Potential contamination of human food can occur when the contaminated flies land on clean food or food preparation surfaces. Filth flies are generally prolific breeders, and a single food source can yield thousands of flies per week. Large numbers of flies can be annoying and psychologically stressful for some people. They also have the potential to contaminate body sores and open wounds, as well as food. Laboratory studies show filth flies can pick up over 1000 different pathogens, but only a few are 'real world' risks, primarily causing diarrhoea and dysentery. Filth flies can fly long distances, potentially carrying pathogens from remote areas such as garbage dumps, sewage lagoons, animal feedlots and animal carcasses. Appendix 7 summarises the pathogens filth flies transmit.

The primary filth fly families are Muscidae (muscids), Calliphoridae (blow flies, bottle flies) and Sarcophagidae (flesh flies).

Muscids
(Family Muscidae)

Muscids are the primary group of nuisance filth flies affecting people. The house fly (*Musca domestica*), found worldwide, is a dull grey fly, 7 to 10 mm (0.17 to 0.25 inch) long with four dark, narrow dorsal longitudinal thoracic stripes. It is a strong flier, flying up to 32 km (20 miles), but normally less than 3 km (<2 miles). House flies lay eggs on various organic materials, includ-

House fly (*Musca domestica*). Source: Stephen L. Doggett

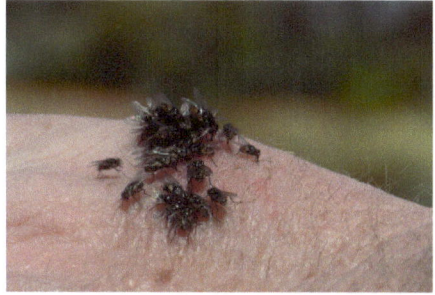

Australian bush flies (*Musca vetustissima*). Source: John Jennings/Flickr. CC BY 2.0. Available from https://www.flickr.com/photos/124930081@N08/14843601919/

ing manure, carrion, kitchen refuse, garbage, cesspool material, decaying fruits or vegetables, and most human foods. They produce large larval masses so you can encounter huge, annoying swarms both indoors and outdoors. They likely transmit numerous pathogens, especially diarrhoeal agents.

The *Musca sorbens* complex species (*Musca biseta*, dog dung fly; *Musca sorbens*, eye fly/bazaar fly; *Musca vetustissima*, Australian bush fly) are attracted to and feed on wounds, ulcers, and so on, but they also feed on mucous secretions around the mouth and eyes of humans and other animals, creating a major nuisance. They appear similar to the house fly, but have two dark, broad longitudinal thoracic stripes. Sometimes they are the most common filth flies in hot, dry regions within their respective ranges. All of these flies, especially *M. sorbens*, are especially problematic in refugee camps or other concentrated, malnourished populations having poor sanitation, and they can be a major management problem for disaster relief efforts. Because the flies are not easily deterred by brushing them aside, they return to rest on people's faces and sometimes in large numbers. When this

happens, people weakened by malnourishment or disease often cease to combat the flies where they feed on fluids from the mouth, eyes and nostrils. This may result in epidemic conjunctivitis or trachoma (*Chlamydia trachomatis*) in such populations, which can cause blindness if not properly treated. *Musca sorbens* also feed on faeces, carrion and garbage, and, as such, they can mechanically transmit several enteric pathogens.

The face fly (*Musca autumnalis*) can be a significant nuisance pest near pastures, barnyards and other livestock operations. They resemble a house fly in appearance,

Bazaar fly (*Musca sorbens*). Source: Muhammad Mahdi Karim/Wikipedia. GFDL 1.2. Available from https://en.wikipedia.org/wiki/Musca_sorbens#/media/File:Musca_sorbens.jpg

Face fly (*Musca autumnalis*). Source: Janet Graham/ Flickr. CC BY 2.0. Available from https://www.flickr. com/photos/130093583@N04/17648436895/

Green bottle flies (*Lucilia sericata*). Source: Giles Gonthier/Flickr. CC BY 2.0. Available from https:// www.flickr.com/photos/gillesgonthier/2495235129/

although they are larger. Similar to the house fly, the thorax has four dark, longitudinal stripes. To the layperson, these flies are difficult to distinguish from house flies. They are distributed throughout Europe, including Great Britain, western Siberia through much of Asia, North Africa, the Middle East, and North America from southern Canada southward. They most commonly breed in livestock manure. Females aggressively swarm around the face of livestock and people during daylight hours. Populations are generally largest during late summer in temperate regions.

Blow flies and bottle flies (Family Calliphoridae)

Over 1000 species of calliphorids occur around the world. These flies are the same size or slightly larger than house flies. Their brightly coloured bodies easily distinguish them, and they can be shiny black, blue, green, copper or bronze. Their preferred food is carrion, but they are also attracted to meat-laden garbage, and wounds (see 'Myiasis flies'). The presence of calliphorids in large numbers usually indicates an animal has died nearby. They deposit eggs on carrion or other suitable

media immediately or up to 2 days after death depending on the species, the season and situation. Most calliphorids are associated with temperate and tropical climates, but *Phormia regina*, the black blow fly, is also found throughout cooler, higher elevation regions worldwide.

Flesh flies (Family Sarcophagidae, Sarcophaga spp.)

This is a large family of over 2500 species. Similar to house flies, most adult sarcophagids are dark-coloured (grey or black) with three dark thoracic stripes, but

Flesh fly (*Sarcophaga* sp.). Source: Gail Hampshire/ Flickr. CC BY 2.0. Available from https://www.flickr. com/photos/gails_pictures/33897341702/

Old World latrine fly (*Chrysomya megacephala*). Source: Bob Peterson. CC BY 2.0. Available from https://www.flickr.com/photos/pondapple/9089189662/

they are slightly larger and have a distinctly chequerboard pattern on the abdomen. Flesh flies usually seek carrion, meat scraps or decaying vegetable matter. Instead of laying eggs, the females are larviparous, giving birth to hatched larvae. The Old World latrine fly, *Chrysomya megacephala,* is a common Indo-Australian area pest and has been introduced to the Afrotropical and Neotropical regions.

Large numbers of sarcophagids may indicate that a dead animal is nearby.

Filth fly prevention and treatment

Situational awareness, avoidance and sanitation are your best safeguards against filth flies. Large filth fly populations denote a likely sanitation issue, as well as potential disease transmission risk. Appropriate personal protection measures around agricultural settings or densely populated sites with poor sanitation (i.e. refugee, disaster recovery or transient camps) are recommended to reduce contact with these flies. Eating and sleeping within screened (14- to 16-mesh) and/or sealed, air-conditioned facilities, and keeping doors, windows and vents closed helps exclude filth flies. Eliminating breeding sites (decomposing organic material, carrion, manure, compost pile, sewage lagoons, latrines, etc.) from around human habitations helps reduce fly densities. During warmer periods, garbage should be stored in an approved refuse disposal container some distance away from dwellings, buried (under >30 cm or 1 foot of soil) or burned.

Fleas
(Order Siphonaptera)

All adult fleas are temporary, obligate, parasites of warm-blooded animals. There are ~2500 species in 200 genera of fleas inhabiting every continent, including the Arctic and Antarctica. The majority of species parasitise mammals (94%), while others parasitise birds (6%). Pest fleas can be artificially divided into two broad types based on the adult female behaviour: (1) actively moving fleas; and (2) attached fleas. Attached adult female fleas either remain attached to the surface or invade the skin. Actively moving fleas move about the host, briefly attaching to feed then releasing several times a day. The latter readily leave their host, opportunistically transferring to other available hosts. When flea populations are large, their biting of humans and pets can be especially annoying and distracting. Flea bites also can result in transmission of serious diseases.

The anatomy and behaviour of fleas make them formidable pests and disease vectors. Females are larger than males, but both sexes are small (1–4 mm or 0.04–0.16 inches long), wingless, brownish in colour and laterally flattened. Fleas are well known for their ability to jump up to 30 cm or 1 foot to get on a host. The lateral body compression facilitates their movement among body hairs or feathers of the host. Moreover, their bodies have numerous bristles that are pointed towards their posterior end to aid forward movement between the hairs, and to catch on hairs, thus impeding a host's efforts to dislodge them. Their piercing-sucking mouthparts and salivary fluids permit blood feeding, and their bites are annoying and painful. Following initial pain, bites may produce a reddish-coloured papule that may progress to a pustule, which potentially may become infected. Numerous bites may produce itching and hive-like allergic reactions in some hosts.

Flea biology

All active fleas have similar life cycles but the duration is highly variable, ranging from 18 days to 20 months (usually 30–75 days) or more depending on the species and environmental conditions (e.g. temperature, humidity). They may produce multiple generations per year depending on optimal conditions. A single female flea may produce up to 500 eggs, and adults can be long lived (up to 18 months), which can lead to explosive population growth. The eggs and larvae are often found among animal bedding, carpets or in cracks and crevices of shelters or houses. Larvae feed on organic debris (e.g. dried blood, scales, dander, faeces, dead fleas) and they are generally found in the same area where their hosts occur, such as bedding. When adult fleas emerge from the pupal cocoon, they usually stay quiescent until host movements, vibrations, warmth, and so on, stimulate them to jump on the host and begin feeding. Newly moulted adults of some species remain (in a pre-emergent state) inside their pupal cocoon for several months until stimulated by nearby host vibrations, warmth and scents stimulate them to break out and jump onto a host. They can survive without feeding for 150 days or more depending on environmental conditions (temperature and humidity), while they await a host to return. It is common for people to leave their home for a few weeks only to return home to an enormous, hungry adult flea population waiting to attack.

Many fleas are not host specific, feeding multiple times on various animals including people, thus increasing their potential to transmit disease agents. In addition, they frequently leave their hosts, wandering about the substrate (e.g. bedding, nest, blanket) between blood meals. Combined, these various traits make fleas important disease vectors in many areas of the world. Fleas are competent vectors of bubonic plague, endemic typhus (flea-borne or murine typhus) and tularemia.

Plague

The last worldwide plague outbreak occurred during the late 19th century but several countries still report plague cases each year. Fleas acquire plague bacteria (*Yersinia pestis*) while feeding on infected rodents. Some 200 rodent species harbour plague and there are numerous infected wild rodent hot spots each year around the world including the United States (west of the 100th meridian), Peru, Bolivia, Brazil, Ecuador, India, and five African and six Asian countries. When an infected animal dies and its body begins cooling, the fleas immediately leave to find the nearest warm-blooded animal. This behaviour can lead to transmission of plague bacteria.

The plague bacteria actually blocks the fleas' digestive tract, preventing ingestion, which results in the flea regurgitating when it tries to feed. Therefore, the infected and starving flea may jump from host to host, biting more and more frequently and regurgitating plague bacteria into the bite wounds of these hosts. Should the transfer of the plague bacteria be successful, the first symptoms appear in 3 to 4 days and include a sudden fever, chills, weakness, malaise and headache. Usually by the second day, the characteristic painful lymph node swellings called buboes can be detected. The most common bubo sites are the groin, armpit and neck lymph nodes. Left untreated, these erupt forming open ulcers, which slowly heal. Antibiotics can easily treat plague but, left untreated, plague can be fatal. There is a vaccine available that offers partial protection to those needing to go to plague outbreak areas (e.g. relief efforts).

Endemic typhus

Fleas also transmit the bacteria that cause endemic typhus (*Rickettsia typhi*) in their faeces. Rubbing contaminated faeces into itchy flea bites, mucous membranes (eyes, nose and mouth) and skin lesions, or accidentally ingesting infected fleas or their faeces can result in infection. The roof, ship or black rat (*Rattus rattus*) and the Norway rat (*Rattus norvegicus*) are reservoirs for endemic typhus worldwide, although primarily in the tropics. Endemic typhus is a relatively mild disease. Here again, the initial symptoms are sudden fever, chills, weakness, malaise, headache and body rash, but antibiotics readily control the disease. Although rare, endemic typhus can be fatal if untreated, primarily among those over 50 or in individuals having compromised immune systems.

Tularemia

Another bacterial pathogen (*Francisella tularensis*) causes tularemia (rabbit fever, deer-fly fever, Oharra's disease, yatobyo or lemming fever). The bacterium naturally infects ~100 species of wild rodents, 50 arthropods, and various other animal species throughout the Northern Hemisphere between latitude 37° and 71° north and it can be transmitted by various ticks, biting flies (*Chrysops*), and fleas. Cases of tularemia have been reported from North America, portions of Europe, Scandinavia and Asia, including Japan. Symptoms usually begin within hours, but they may take 21 days or more to manifest. The abrupt onset symptoms may include fever, malaise, headache, backache, severe sore throat, nausea, abdominal pain, vomiting, diarrhoea, and weakness progressing to anorexia, stupor, delirium, stiff neck and rarely death. Tularemia usually produces a marked reaction where it enters the body whether it is the bite site, eye, mucous membrane or lung tissue. Characteristic necrotic lesions or ulcers are usually present on the skin. Antibiotics readily treat tularemia infections.

Cat and dog fleas

The common cat flea (*Ctenocephalides felis*) and dog flea (*Ctenocephalides canis*) are cosmopolitan pests. Cat fleas are typically more abundant and broadly distributed than dog fleas. These fleas attack various mammals including foxes, raccoons, rats, cats, dogs, and humans. Large, cat and dog flea populations can inflict numerous bites and potentially

Cat flea (*Ctenocephalides felis*). Source: David E. Bowles

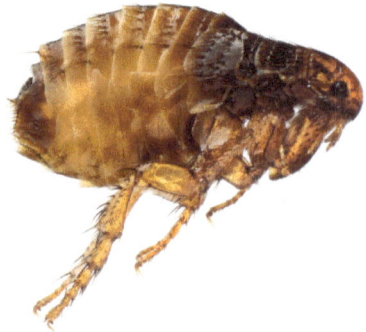

Dog flea (*Ctenocephalides canis*). Source: Stephen L. Doggett

cause a dermatitis requiring medical treatment (see 'Flea prevention and treatment'). Both species will breed outdoors during the summer in vacant lots, under houses, in barn, and similar locations, particularly if stray dogs or cats are about.

Human flea

The human flea (*Pulex irritans*) also has a cosmopolitan distribution. It infests a wide variety of mammals including domestic rats, pigs, cats, dogs, coyotes, goats, prairie dogs, ground squirrels, burrowing owls, skunks, badgers and humans. It can

Oriental rat flea (*Xenopsylla cheopis*). Source: Stephen L. Doggett

Human flea (*Pulex irritans*). Source: Stephen L. Doggett

transmit plague bacteria under laboratory conditions and may play a role in transmitting that pathogen in the wild. It is an important flea species attacking people along the Pacific Coast of North America, often causing dermatitis or allergic reactions when its populations are large. Occasionally it becomes abundant on farms, particularly in abandoned pigpens.

Oriental rat flea

Oriental rat (tropical- or black-) flea (*Xenopsylla cheopis*) normally infests Norway and roof rats in urban settings worldwide. The oriental rat flea is the primary vector of bubonic plague and endemic (flea-borne or murine) typhus. Other species in the genus *Xenopsylla* also transmit these two diseases, including *X. brazilinesis* in South America, India and Africa (Uganda, Kenya and Nigeria), and *X. astia* in India and Myanmar.

Northern rat flea

Northern rat flea or brown rat flea (*Nosopsyllus fasciatus*) commonly infests the domestic Norway rat, roof rat, the house mouse (*Mus musculus*) and other small mammals throughout temperate areas worldwide. Although this species does not readily bite people, it may be a zoonotic plague vector.

European mouse flea

European mouse flea (*Leptopsylla segnis*) is cosmopolitan in distribution, and it frequently is a pest indoors. It has been introduced to many areas outside its native range via infested rats and mice that entered through coastal ship ports. Mouse fleas can transmit endemic typhus, but they are considered poor vectors of that pathogen. They are not natural plague vectors, but laboratory studies show they can be infected with plague bacteria.

Squirrel flea

Squirrel flea (*Orchopeas howardii*) commonly infest various squirrels and birds from eastern Canada to Central America. They sometimes become serious household pests when squirrels invade home attics or crawl spaces to nest. Fleas breeding in attic nest material may subsequently invade other areas of a house

Sticktight flea (*Echidnophaga gallinacea*). Source: Stephen L. Doggett

where they attack people. Squirrel fleas are potential plague vectors.

Sticktight flea

Sticktight flea (*Echidnophaga gallinacea*) is a small species that attacks rats, cats, dogs, rabbits, ground squirrels, horses, fowl and many other animals, including people. Although native to the Americas, they are now essentially distributed worldwide. Sticktight fleas get their name because they firmly attach to the skin, often forming skin ulcers due to their constant feeding. Eggs are laid and hatch in these ulcers. The larvae drop to the ground to feed on organic matter until they pupate and emerge to complete the life cycle. All life stages of this flea were historically abundant among poultry yards and adjacent buildings. Sticktight fleas naturally carry both plague and endemic typhus, but they only play a minor disease transmission role because the females remain permanently attached.

Hen fleas

The western chicken (hen) flea (*Cerato-phyllus niger*) and the European chicken (hen) flea (*Ceratophyllus gallinae*) are particularly large fleas that attach to their hosts, and they have painful bites. The former species is distributed in western North America, while the latter species has a worldwide distribution. Both occasionally become extremely abundant among chicken coops and wild bird nests. Large numbers occasionally attack people.

Tunga flea

Tunga flea (*Tunga penetrans*) or jigger (chigger, chigo, chique or sand flea) is widely distributed in tropical and subtropical regions of Africa, India, the Caribbean, and North and South America. Among the fleas that attach to their hosts, this species is by far the most annoying and medically important. The principle hosts of this flea are pigs and humans, but they also infest dogs, cats, rats and some domestic fowl. After emerging from the pupal stage, the biology of the tiny (1 mm long or 0.03 inch) male and female jigger is somewhat similar to that of the actively moving fleas. After mating, the males remain free roaming and feed on whatever

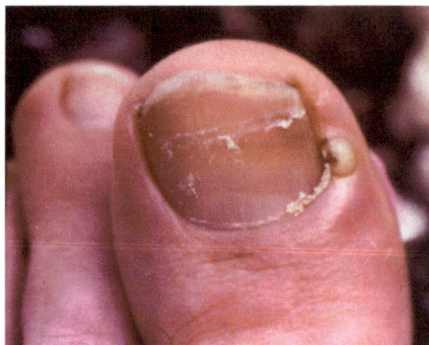

Chigoe (*Tunga penetrans*) embedded in a toe. Source: D.R. Roberts, US Department of Defense, Uniformed Services University of the Health Sciences

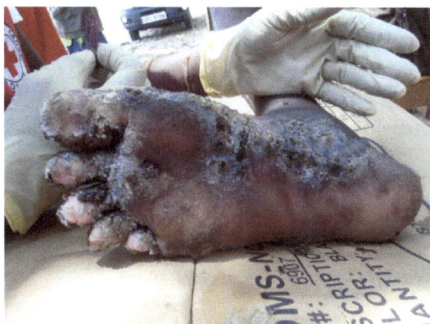

A foot heavily infested by chigoes (*Tunga penetrans*): Source: R. Schuster/Wikipedia. CC BY-SA 3.0. Available from https://en.wikipedia.org/wiki/Tunga_penetrans#/media/File:Jigger_infested_foot_(2).jpg

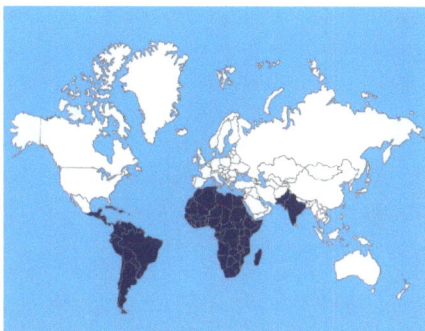

Distribution of Chigoes (*Tunga penetrans*).

host might be available. However, the fertilised females seek a host where upon they firmly attach to the skin, usually in a crack or hidden area. Most commonly, they attach to the host where they embed themselves between the toes, soles of the feet or under toenails, but occasionally they can be located elsewhere on the body, including the genitalia, perianal area and legs. The fertilised eggs in the female quickly grow, expanding her abdomen. After 8 to 12 days, the flea becomes enveloped in the host's tissue, resulting in an irritating, excruciatingly painful sore. The wound appears as a pea size (10 mm, 0.4 inch), pale capsule with a central dark spot, which is the hole used for respiration. Pus fills the capsule, which subsequently ulcerates forming a wound up to 2 cm (0.8 inch) in diameter. This inflamed ulcer not only stimulates scratching that expels and may disperse the eggs, it may lead to secondary infection, gangrene or tetanus if left untreated. Some severe cases have required amputation, and death due to infection has occurred, albeit rarely. The female remains embedded in the host's tissue where she keeps expelling eggs for around 2 weeks or until she dies. The slightly sticky, expelled eggs tend to adhere to surrounding substrate (sloughed skin cells, hairs, dust, fibres, etc.).

Chigoe (*Tunga penetrans*) removed from a patient. Source: Puce-chique/Wikipedia. Public domain. Available from https://en.wikipedia.org/wiki/Tungiasis#/media/File:Puce_chique_(Tonga_penetrans).jpg

Flea prevention and treatment

Using personal protection measures in combination with situational awareness is the best means of avoiding contact with fleas. Avoid and do not linger around wild animal nests or borrows, or cages or pens of domestic animals and birds. Rodent and bird infested buildings should also be avoided because they will likely have flea populations as well. Within the range of tunga fleas, always wear closed toed shoes, thoroughly wash feet at least once per day, and avoid sleeping on bare, especially sandy, ground that might harbour these fleas.

If primitive housing such as cabins or campsites is suspect for fleas and other pests, use of a treated bed net and sleeping elevated is highly recommended. Should fleas infest a domestic dwelling, the floors/carpet should be swept (vacuumed) and/or cleaned frequently until the population disappears. Such cleaning helps eliminate all life stages, especially eggs, larvae and pupae. Indoor and outdoor pets should be provided flea treatments based on a veterinarians guidance, and periodically washing pets is recommended. Stray animals such as cats or dogs and wild animals such as squirrels should not be allowed to bed or nest around, under or in

homes. Should your home develop a flea infestation that cannot be controlled using this guidance, a licenced pest controller should be consulted.

For flea bites, topical corticosteroids and/or systemic antihistamines are useful for reducing itching and localised dermatitis. In severe cases, or if secondary infections occur, professional medical treatment should be obtained. Similarly, any unexplained illness following known contact with fleas, recently dead animals, or animal burrows, should be cause for concern and medical attention should be sought at the earliest opportunity. In the absence of medical assistance, tunga fleas can be removed using a sharp, sterile needle or blade that can gently tease out the whole tunga flea. However, the resulting wound should be thoroughly cleaned and sterilised with alcohol or other appropriate disinfectant and covered with an antiseptic dressing to prevent infection.

Selected references

Acha PN, Boris S (1987) *Zoonoses and Communicable Diseases Common to Man and Animals.* 2nd edn. Pan American Health Organization, Washington DC, USA.

Akre RD, Greene A, MacDonald JF, Landolt PJ, Davis HG (1980) *Yellowjackets of America North of Mexico.* Agriculture Handbook No. 552, U.S. Department of Agriculture, Washington DC, USA.

Alpern JD, Dunlop SJ, Dolan BJ, Stauffer WM, Boulware DR (2016) Personal protection measures against mosquitoes, ticks, and other arthropods. *The Medical Clinics of North America* **100**, 303–316. doi:10.1016/j.mcna.2015.08.019

Barnes RD (1987) *Invertebrate Zoology.* 5th edn. Saunders College Publishing, New York, USA.

Beccaloni J (2009) *Arachnids.* University of California Press, Berkeley CA, USA.

Benenson AS (1990) *Control of Communicable Diseases in Man.* 75th edn. American Public Health Association, Washington DC, USA.

Biery TL (No date) Venomous arthropod handbook, envenomation, symptoms/treatment, identification biology and control. Air Force Pamphlet 161–43, United States Air Force School of Aerospace Medicine, Brooks AFB, TX, USA.

Britton EB (1973) *The Insects of Australia.* Melbourne University Press, Carlton, Victoria.

Bücherl W, Buckley EE (Eds) (1971) *Venomous Animals and Their Venoms. Volume III. Venomous Invertebrates.* Academic Press, New York, USA.

Bush SP, King BO, Norris RL, Stockwell SA (2001) Centipede envenomation. *Wilderness & Environmental Medicine* **12**, 93–99. doi:10.1580/1080-6032(2001)012[0093: CE]2.0.CO;2

De La Torre-Bueno JR (1973) *A Glossary of Entomology.* New York Entomological Society, New York, USA.

Debboun M, Frances S, Strickman D (Eds) (2007) *Insect Repellents Handbook.* 2nd edn. CRC Press, Boca Raton FL, USA.

Dzelalija B, Medic A (2003) *Latrodectus* bites in Northern Dalmatia, Croatia: clinical, laboratory, epidemiological, and therapeutical aspects. *Croatian Medical Journal* **44**, 135–138.

Feldmeir H, Eisle M, Saböia-Moura RC, Heukelbach J (2003) Severe tungiasis in underprivileged communities: a case series from Brazil. *Emerging Infectious Diseases* **9**, 1–7.

Frazier CA, Brown FK (1980) *Insects and Allergy and What to Do about Them.* University of Oklahoma Press, Norman, OK, USA.

Garb JE, González A, Gillespie RG (2004) The black widow spider genus *Latrodectus* (Aranae: Theridiidae): phylogeny, biogeography, and invasion history. *Molecular Phylogenetics and Evolution* **31**, 1127–1142. doi:10.1016/j.ympev.2003.10.012

Gershwin L-A (2013) *Stung! On Jellyfish Blooms and the Future of the Ocean.*

University of Chicago Press, Chicago, IL, USA.

Gershwin L-A (2016) *Jellyfish: A Natural History*. University of Chicago Press, Chicago, IL, USA.

Goddard J (1989) *Ticks and Tickborne Diseases Affecting Military Personnel*. USAF School of Aerospace Medicine, USAF-SAM-SR-89-2, Brooks Air Force Base, TX, USA.

Goddard J (1993) *Physicians Guide to Arthropods of Medical Importance*. CRC Press, London, UK.

Goddard J (1994) Direct injury from arthropods. *Laboratory Medicine* **25**, 365–371. doi:10.1093/labmed/25.6.365

Goddard J (2000) *Infectious Diseases and Arthropods*. Humana Press, Totowa, NJ, USA.

Goddard J (2013) *Physicians Guide to Arthropods of Medical Importance*. 6th edn. CRC Press, Boca Raton, FL, USA.

Gorham JR, Rheney TB (1968) Envenomation by the spiders *Chiracanthium inclusium* and *Argiope aurantia*. *Journal of the American Medical Association* **206**, 1958–1962. doi:10.1001/jama.1968.03150090034007

Grundy JH (1981) *Arthropods of Medical Importance*. Noble Books Limited, Chilbolton, Hampshire, UK.

Halstead BW (1965) *Poisonous and Venomous Marine Animals of the World. Volume One – Invertebrates*. United States Government Printing Office, Washington DC, USA.

Halstead BW (1995) *Dangerous Marine Animals that Bite, Sting, Shock, or Are Nonedible*. 3rd edn. Cornell Maritime Press, Centerville, MD, USA.

Harwood RF, James MT (1979) *Entomology in Human and Animal Health*. 7th edn. Macmillan Publishing, New York, USA.

Hawkeswood TJ (2003) *Spiders of Australia: An Introduction to their Classification, Biology and Distribution*. Series Faunistica No. 31, Pensoft Publishers, Sofia, Bulgaria.

Hölldobler B, Wilson EO (1990) *The Ants*. Belknap Press of Harvard University Press, Cambridge, MA, USA.

James WD (Ed.) (1994) *Textbook of Military Medicine, Part III, Disease and the Environment, Military Dermatology*. Office of the Surgeon General, Walter Reed Army Medical Center, Washington DC, USA.

Kettle DS (1995) *Medical and Veterinary Entomology*. 2nd edn. CABI Publishing, Oxford, UK.

Koh JKH, Ming LT (2014) *Spiders of Borneo: with Special Reference to Brunei*. Opus Publications, Kota Kinabalu, Sabah, Malaysia.

Lago PK, Goddard J (1994) Identification of medically important arthropods. *Laboratory Medicine* **25**, 298–305. doi:10.1093/labmed/25.5.298

Mann KH, GA Kerkut (2013) *Leeches (Hirudinea); Their Structure, Physiology, Ecology, and Embryology*. Elsevier Science, Amsterdam, Netherlands.

Manson-Bahr PEC, Apted FIC (1982) *Manson's Tropical Diseases*. 18th edn. Bailliere Tindall, London, UK.

McGoldrick J, Marx JA (1992) Marine envenomations part 2: invertebrates. *The Journal of Emergency Medicine* **10**, 71–77. doi:10.1016/0736-4679(92)90014-K

McHugh CP (1994) Arthropods: vectors of disease agents. *Laboratory Medicine* **25**, 429–437. doi:10.1093/labmed/25.7.429

Mebs D (2002) *Venomous and Poisonous Animals: A Handbook for Biologists, Toxicologists and Toxinologists, Physicians and Pharmacists*. CRC Press, Boca Raton, FL, USA.

Mullen GR, Durden LA (Eds) (2009) *Medical and Veterinary Entomology*. 2nd edn. Academic Press, Orlando, FL, USA.

Norris R (2017) Centipede envenomations. *eMedicine Journal* **3**.

Peters W (1992) *A Colour Atlas of Arthropods in Clinical Medicine*. Wolfe Publishing, London, UK.

Schmidt JO (2016) *The Sting of the Wild*. Johns Hopkins University Press, Baltimore, MD, USA.

Schofield CJ (1994) *Triatominae, Biology and Control*. Eurocommunica Publications, Bognor Regis, UK.

Shannon E, Bruzelius N (Eds)(2013) *EWG's Guide to Better Bug Repellents*. Environmental Working Group, Washington DC, USA.

Smith RL (1982) *Venomous Animals of Arizona*. Cooperative Extension Service, University of Arizona, Tucson, AZ, USA.

Southcott RV (1970) Human injuries from invertebrate animals in the Australian seas. *Clinical Toxicology* **3**, 617–636. doi:10.3109/15563657008990136

Strickland GT (1984) *Hunter's Tropical Medicine*. 6th edn. W.B. Saunders Company, Philadelphia, PA, USA.

Teyssie F (2015) *Tarantulas of the World: Theraphosidae*. N.A.P. Editions, Rodez, France, <www.napeditions.com>.

Triplehorn CA, Johnson NF (2005) *Borror and DeLong's Introduction to the Study of Insects*. 7th edn. Thomson, Brooks/Cole, Belmont, CA, USA.

Tyagi BK (2003) *Medical Entomology: A Handbook of Medically Important Insects and Other Arthropods*. Vedams eBooks (P), New Delhi, India.

United States Air Force (2002) *Guide to Surveillance of Medically Important Vectors and Pests*. USAF Force Protection Battlelab, Lackland Air Force Base, TX, USA.

United States Centers for Disease Control and Prevention (No date) *Pictorial Keys to Arthropods, Reptiles, Birds and Mammals of Public Health Significance*. U.S. Department of Health and Human Services, Centers for Disease Control and Prevention, Atlanta, GA, USA.

Vetter RS (2008) Spiders of the genus *Loxosceles* (Araneae, Sicariidae): a review of biological, medical and psychological aspects regarding envenomations. *The Journal of Arachnology* **36**, 150–163. doi:10.1636/RSt08-06.1

Vetter R (2015) *The Brown Recluse Spider*. Cornell University Press, Ithaca, New York, USA.

Warren KS, Mahoud AF (1984) *Tropical and Geographical Medicine*. McGraw-Hill Book Company, New York, USA.

Webb C, Doggett S, Russell R (2016) *A Guide to Mosquitoes of Australia*. CSIRO Publishing, Melbourne.

Wong S (Ed.) (2016) The medical letter on drugs and therapeutics *Insect repellents* **58** (1498), 83.

Ythier E (2010) *Scorpions of the World*. N.A.P. Editions, Rodez, France, <www.napeditions.com>.

Zumpt F (1965) *Myiasis in Man and Animals in the Old World*. Butterworths, London, UK.

Web-based sources of hazardous invertebrate references and information

Australian Government Smart Traveller website: www.smartraveller.gov.au

American Arachnological Society: www.americanarachnology.org

Brown recluse spider information: spiders.ucr.edu/

Euscorpius, the on-line publication on Scorpion taxonomy, etc.: www.science.marshall.edu/fet/euscorpius

The Armed Forces Pest Management Board LHD: www.acq.osd.mil/eie/afpmb/livinghazards.html

The on-line Journal of Venomous Animal Toxins, Botucatu, Brazil: www.scielo.br/scielo.php

The Scorpion Files: www.ub.ntnu.no/scorpion-files/

Toxinology website, Adelaide, Australia: www.toxinology.com

University of Michigan Museum of Zoology Diversity Web: animaldiversity.ummz.umich.edu/

U.S. Centers for Disease Control and Prevention (CDC): www.cdc.gov

Walter Reed Biosystematics Unit: www.wrbu.org

World Health Organization (WHO): www.who.org

Glossary

anaphylaxis – an increased sensitivity to a foreign compound so that a second exposure brings about a severe reaction sufficient to induce shock and death.

analgesic – a chemical used to relieve pain without producing anaesthesia or loss of consciousness.

anaemia – low red blood cell or haemoglobin concentration.

anaesthetic – a drug that causes reversible numbness.

antihistamine – a chemical used to counteract histamine (a normal body chemical) related reactions to various allergens.

arachnidism – a spider bite.

arbovirus – arthropod-transmitted viruses.

arrhythmia– abnormal heartbeat or rhythm.

arthralgia – pain in joints (ankles, knees, elbows, wrists, shoulders).

auricular – the sense of or related to the sense of hearing.

bradycardia – abnormally slow heart rate.

capitulum – the head-end of a tick.

CDC – United States Center for Disease Control and Prevention.

cephalothorax – head and thorax combined into one structure (body region).

chitin – a colourless epidermal secreted nitrogenous polysaccharide intermediate between proteins and carbohydrates that makes up the hard arthropod exoskeleton.

corticosteroid – natural hormones produced by vertebrates and their synthetic analogues, which work as anti-inflammatory agents.

conjunctivitis – inflammation of the mucous membranes lining the eyelids and covering the anterior surface of the eyeball.

coxal – referring to the first or basal segment or coxa of an insect leg.

cutaneous – referring to the skin.

cyanosis – bluish, slate-like or dark purple skin colour due to excessive concentration of reduced haemoglobin in the blood.

cytotoxic – poisoning of cells.

cuticle – the arthropod exoskeleton's outer most layer composed of chitin.

debridement – the surgical removal of lacerated, devitalised or contaminated tissue.

delirium – a temporary disorientation, usually accompanied by illusions and hallucinations.

delusion – a belief that is steadfastly maintained despite overwhelming and widely accepted evidence to the contrary.

dermatitis – inflammation of the skin.

dialysis – a medical procedure using rates at which substances diffuse through a semipermeable membrane to remove wastes or toxins from the blood while adjusting fluid and electrolyte imbalances.

disseminated intravascular coagulation – a complex and controversial systemic thrombohaemorrhagic disorder involving the generation of intravascular fibrin and the consumption of procoagulants and platelets.

diurnal – occurring or active during the daylight hours.

dorsoventrally – extending along or denoting an axis joining the dorsal and ventral surfaces.

ectoparasite – a parasite that lives on the outside of its host.

elytra – hardened forewings of beetles (Order Coleoptera).

encephalitis – inflammation of the brain.

entomophobia – an irrational fear of insects and their relatives, or the damage or diseases they are capable of inflicting.

envenomation – bite and/or sting venom injection into the body.

EPA – United States Environmental Protection Agency.

erythema – unusual redness of the skin caused by capillary congestion, resulting from inflammation, as in heat or sunburn.

erucism – envenomation by moths and caterpillars (Order Lepidoptera).

eschar – a sloughing or crusting, which forms on the skin after the tissue dies.

facultative parasite – exhibiting an indicated lifestyle under some environmental conditions but not under others.

fibrin – an insoluble protein essential to blood clotting.

gastrointestinal – of, relating to, or affecting both stomach and intestine.

Haller's organ – hairy olfactory sensory organ found on body, legs and mouthparts of ticks.

hallucination – perception of an imaginary physical object or a mistaken impression (notion).

haematoma – a swelling of blood, which occurs in an organ or tissue resulting from ruptured blood vessels.

haemolytic – relating to the rupture or disintegration of red blood cells.

haemorrhage – the escape or release of blood from a ruptured vessel.

histamine – a human immune response chemical causing allergic reactions via blood vessel enlargement, airway tightening and increased heart rate (decreased blood pressure).

hives – a skin rash with red, raised, itchy bumps, which may also burn or sting. Also known as urticaria.

hyperpyrexia – extremely high fever.

hyperreflexia – overactive reflexes.

hypertension – persistently high arterial blood pressure.

hypostome – ticks' mouthpart containing sensory organs, cutting organs, and an attachment/feeding organ.

hypotension – abnormally low blood pressure.

immunosuppressive – partially or completely suppressing the immune response.

jaundice – yellowing of the skin or whites of the eyes.

larviparous – depositing living larvae, instead of eggs.

latrodectism – injury or illness caused by the bite of various spiders in the genus Latrodectus spiders (black widow spiders and related species).

lepidopterism – a disease that is caused by butterflies, moths and their caterpillars (Order Lepidoptera).

maculopapular – a patch of skin that is altered in colour and usually elevated that is a characteristic feature of various diseases.

malaise – a general feeling of discomfort, illness or uneasiness.

meningitis – inflammation of the brain meninges (outer lining).

moult – to cast off the outgrown exoskeleton (skin, cuticle) to grow and transform from one life stage to the next (e.g. larvae to nymph or larvae to papa, larvae to nymph, nymph to adult).

mydriasis – dilation of the pupils.

myiasis – invasion of the tissue of man or animals with the larvae (maggots) of certain flies (Order Diptera) that consume flesh or body fluids for sustenance.

neurotoxic – toxic to the nerves or nervous tissue.

neurotoxin – toxins that primarily affect the nervous system of an animal.

necrosis – death of cells, tissue or bone by enzymatic degradation of surrounding healthy tissues.

necrotising arachnidism syndrome – a response to necrotic arachnid venom where living tissue is rapidly destroyed around the bite site.

obligate parasite – a parasite completely dependent on its host.

ocular – referring to the eyes.

otitis – ear inflammation.

otoacariasis – infestation of the ear with mites.

PAHO – Pan-American Health Organization.

palpable – something that can be touched or felt.

palps – sensory appendages on arthropod modified mouthparts and mollusc head organs.

papules – small elevated skin lesions.

paranoia – mental state characterised by unwarranted belief in persecution or feelings of grandeur usually without hallucinations.

paresthesia – an abnormal tactile sensation, often described as creeping, burning, tingling or numbness.

parasitosis – infestation with or disease caused by parasites.

pathogen – microorganisms (viruses, bacteria, etc.) that cause disease.

perianal – the anus and surrounding area.

pheromone – chemical secreted or excreted that stimulates a behavioural response in members of the same species.

piloerection – involuntary erection or bristling of hairs due to a sympathetic reflex usually triggered by cold, shock, or fright or due to a sympathomimetic agent.

ppt – parts per thousand as units for expression of concentrations of dilute solutions.

proboscis – a slender, tubular invertebrate sucking or piercing organ.

prostration – total exhaustion, weakness or collapse.

pruritis – intense, chronic itching.

pseudomyiasis – accidental infestations with fly larvae such as when they are inhaled or swallowed inadvertently with food.

pulmonary oedema – swelling of the lung tissue due to an excessive accumulation of fluid.

pustule – a small inflamed elevated skin area containing pus or a distinct small spot resembling a pimple or blister.

questing – ticks' side-to-side swaying behaviour while holding the first pair of legs out stretched, like antennae, with the claws wide open sensing the environment to grab a host.

sebaceous – resembling or characterised by fat.

seizures – abnormal brain electrical discharge and associated symptoms.

sepsis – a foreign microbe or toxic invasion stimulating a hyperactive natural immune response affecting local or systemic body functions.

setae (seta singular) – stiff hairlike structures or bristles on an invertebrate.

sympathomimetic syndrome – a condition generally characterised by a broad suite of symptoms including: delusions, paranoia, tachycardia or bradycardia, hypertension, hyperpyrexia, diaphoresis (sweating), piloerection, mydriasis, hyperreflexia, seizures, and hypotension with arrhythmias possible in serious cases.

tachycardia – an abnormally rapid heart, usually between 160 and 190 beats per minute.

thrombohaemorrhagic – being haemorrhagic due to lack of effective blood clotting.

toxicognaths – an appendage of arthropods derived from the legs that is capable of injecting venom.

tussockosis – tussock moth caused dermatitis.

urticaria – itchy hives or wheals that are redder or paler than the surrounding area.

urticating – cause a stinging or prickling sensation resulting from contact with certain invertebrate body parts.

urtication – a physiological response to contact with certain invertebrate body parts resulting in a painful burning and itchy skin eruption, or hives, at the point of contact.

urogenital – relating to organs or functions of excretion and reproduction.

vector-borne – carried or transmitted by vectors such as insects or other pests.

WHO – World Health Organization

zoonotic – infectious diseases in nature (usually vertebrates) transmissible to humans.

Appendix 1. United States Environmental Protection Agency registered repellent efficacy and use assessments

Disclaimer: We have only included United States Environmental Protection Agency (EPA) registered repellents, the exceptions being catnip oil and 2-undecanone because those two compounds have limited effectiveness. Although it has limited effectiveness, we included citronella due to its wide availability. We also did not include MGK Repellent 326 (Di-n-propyl isocinchomeronate) or MGK-264 (n-Octyl bicycloheptene dicarboximide, a synergist used to enhance DEET). There are no EPA registered formulations of the MGK compounds and they are not particularly effective. However, the EPA Reregistration Eligibility Decision (RED) states, if properly used, MGK-326 poses no unreasonable adverse human health effects. You can find products containing these active ingredients around the world. The overall effectiveness assessments, advantages and disadvantages columns below refer generically to active ingredient and do not imply quality or performance of any specific products formulated using the active ingredient. We used peer-reviewed scientific literature and EPA labelling registration approvals, reviews and fact sheets for our overall effectiveness assessments estimates (http://cfpub.epa.gov). Actual protection time will vary based on your physiology, activity level and environmental conditions (temperature, humidity, moisture, wind, etc.).

Name	Overall assessment if used appropriately[1,2]	Application site[1]	Most effective concentration[2]	Target pests[1]	Maximum protection duration[1,2]	Advantages[1,2]	Disadvantages[1,2]
DEET[3]	Excellent (best choice)	- skin - fabric - other substrates	33.5%	Biting flies, biting midges, black flies, chiggers, deer flies, fleas, gnats, horse flies, mosquitoes, no-see-ums, sand flies, small flying insects, stable flies and ticks	6–13 h	- longest lasting - broadly effective - most studied - availability	- smells - oily/sticky - plasticiser[9]
Picaridin[4] (Icaridin, KBR 3023)	Excellent (next best choice)	- skin - fabric - other substrates	20.0%	Biting flies, chiggers, fleas, gnats, mosquitoes, no-see-ums, and ticks?	5 h	- odourless - not oily/sticky - efficacy ≡ DEET	- shorter duration - limited data - use restrictions - limited effect on ticks
IR3535[5]	Limited	- skin - fabric - other substrates	20.0%	Biting midges, black flies, deer ticks, gnats, mosquitoes (anopheles weak), no-see-ums, and sand flies	4–6 h	- odourless - not oily/sticky - widely studied in Europe	- shorter duration - use restrictions - plasticiser[9]
Oil of lemon eucalyptus[6]	Limited	- skin - fabric - other substrates	30.0%	Biting flies, gnats, and mosquitoes	3–10 h 3–4 h (Anopheles)	- nicer smell - not oily/sticky	- shorter duration - limited data - use restrictions - safety issues - possible allergens - cannot use in malaria endemic areas
Citronella oil[7]	Poor	- skin - fabric - other substrates	40.0%? limited data	Biting midges, biting flies, black flies, bugs, deer flies, face flies, fleas, flies, flying insects, gnats, horn flies, horse flies, house flies, mosquitoes (limited), 'no-see-ums', stable flies and ticks?	≤1 h	???	- shortest duration - limited data - use restrictions - safety issues - possible allergens - limited effect on ticks - does not repel some mosquitoes such as *Aedes albopictus*

Name	Overall assessment if used appropriately[1,2]	Application site[1]	Most effective concentration[2]	Target pests[1]	Maximum protection duration[1,2]	Advantages[1,2]	Disadvantages[1,2]
Permethrin[8]	Best for intended use	- fabric - other substrates only	40.0%	Broad-spectrum arthropod disease vectors, biting and venomous pests: ants, bees, biting flies, biting midges, bed bugs, black flies, chiggers, deer flies, fleas, gnats, horse flies, lice, mosquitoes, no-see-ums, sand flies, scorpions, small flying insects, spiders, stable flies ticks, wasps, urticating caterpillars	6 wks–years	- lasts and lasts - odourless once dry - broadly effective	- use restrictions

1. Source: US EPA Registration Eligibility Decision (RED) fact sheets
2. Source: Published peer reviewed data
3. N, N diethylmeta-toluamide)
4. 2-(2-hydroxyethyl)-1-piperindinecarboxylic acid 1-methylpropyl ester or KBR 3023, Bayrepel, icaridin
5. Butylacetylaminopropionate (EBAAP)
6. Para-menthane–diol (PMD) or OLE
7. An essential oil composed of over 80 closely related terpenic hydrocarbons, alcohols, and aldehydes compounds (Ceylon type, Java type)
8. (±)-3-Phenoxybenzyl 3-(2, 2-dichlorovinyl)-2, 2 dimethylcyclopropanecarboxylate
9. Plasticiser can harm some synthetic fabrics (rayon and spandex), plastics (watch crystals and some eyeglass frames/lenses), car and furniture finishes.

Appendix 2. United States Environmental Protection Agency registered repellent safety data and safety assessments

Remember: There is no such thing as a harmless substance. Only harmless ways of using a substance! Repellents and pesticides, like any substance, can be harmful if misused! For example, if you drink too much water it can kill you.

Name	Overall risk assessment if used appropriately	Acute oral toxicity LD$_{50}$ [1]	Acute dermal toxicity LD$_{50}$ [1]	Acute inhalation toxicity LC$_{50}$ [1]	Dermal irritation	Eye irritation	Pregnancy and breast feeding [1,2,9]	Age range [1,2] (follow label directions)
DEET[3]	Very low risk - best studied	EPA Cat. III 3664 mg/kg	EPA Cat. III 4280 mg/kg	EPA Cat. IV 5.95 mg/L	EPA Cat. IV None	EPA Cat. III Moderate	1st trimester – avoid (no data) 2nd and 3rd trimester – low risk Breastfeeding – low risk	>2 mos.[10]
Picaridin[4]	Low risk - limited data - limitations	EPA Cat. III 4743 mg/kg	EPA Cat. III >2000 mg/kg	EPA Cat. IV 4.364 mg/L	EPA Cat. IV None	EPA Cat. III Moderate	Avoid	>2 yrs.[11]
IR3535[5]	Low risk - well studied in Europe - limitations	EPA Cat. IV >5000 mg/kg	Per WHO ≤10 /kg - no systemic reactions - erythema observed	EPA Cat. IV >5.10 mg/L	Per WHO No reaction	Per WHO 1.0 mL - mild injury	Avoid	>6 months. Avoid
Oil of lemon eucalyptus[6]	Moderate risk - limited data and limitations	EPA Cat. IV >5000 mg/kg	EPA Cat. IV >5000 mg/kg	EPA Cat. IV >2.17 mg/L	EPA Cat. IV Slight	EPA Cat. II Severe	Avoid	>3 yrs.
Citronella oil[7]	Moderate risk - limited data and limitations	EPA Cat. IV >5000 mg/kg	EPA Cat. III >2000 mg/kg	EPA Cat. IV >5000 mg/ kg	EPA Cat. III Moderate	EPA Cat. III Moderate	Risky, little data available	Avoid

Name	Overall risk assessment if used appropriately	Acute oral toxicity LD$_{50}$[1]	Acute dermal toxicity LD$_{50}$[1]	Acute inhalation toxicity LC$_{50}$[1]	Dermal irritation	Eye irritation	Pregnancy and breast feeding[1,2,9]	Age range[1,2] (follow label directions)
Permethrin[8] NO skin treatment	Very low risk	EPA Cat. III >3580 mg/kg	EPA Cat. III >2000 mg/kg	Per EPA no data	EPA Cat. IV None	EPA Cat. III Moderate	Low risk[12] negligible absorption[13]	>2 yrs. Negligible absorption[13]

1. Lethal dose (LD$_{50}$) or lethal concentration (LC$_{50}$) are levels causing death (via single or limited exposure) in 50% of treated animals. LD$_{50}$ is usually expressed as dose, milligrams (mg), per bodyweight, kilogram (kg). LC$_{50}$ is often expressed as mg dose per volume (e.g. litre (L) of medium (i.e. air or water). The smaller the LD$_{50}$/LC$_{50}$ the higher the toxicity. For comparison, vinegar LD$_{50}$ = 1060 mg/kg and nicotine LD$_{50}$ = 50 mg/kg. LD$_{50}$/LC$_{50}$ do not reflect long-term exposure effects (i.e. cancer, birth defects or reproductive toxicity).
EPA Category classification key:
I = very highly or highly toxic
II = moderately toxic
III = slightly toxic
IV = practically non-toxic
2. Sources: EPA Registration Eligibility Decision (RED) Fact Sheets and published peer reviewed data.
3. N, N diethylmeta-toluamide).
4. 2-(2-hydroxyethyl)-1-piperindinecarboxylic acid 1-methylpropyl ester or KBR 3023, Bayrepel, icaridin.
5. Butylacetylaminopropionate or EBAAP.
6. Para-menthane-diol (PMD) or OLE.
7. An essential oil composed of over 80 closely related terpenic hydrocarbons, alcohols and aldehydes compounds (Ceylon type, Java type).
8. (±)-3-Phenoxybenzyl 3-(2,2-dichlorovinyl)-2,2-dimethylcyclopropanecarboxylate. Permethrin has arthropod repellent properties but is an EPA registered general (public) use pesticide.
9. The CDC considers EPA-registered DEET, picaridin, IR3535, and OLE formulations safe during pregnancy.
10. American Academy of Pediatrics (ACP) recommends using 10–30% DEET formulations on children.
11. Per ACP, 5–10% picaridin formulations can be used on children as an alternative to DEET.
12. Per EPA, no evidence of reproductive effects in pregnant or nursing women or developmental adverse effects in children due to permethrin exposure. Be judicious, consider disease risk.
13. <1/1000 of that from topical permethrin-based lice treatment lotions. Be judicious, consider disease risk.

Appendix 3. Some United States Environmental Protection Agency registered commercial insect repellent product information

Table 1. Some insect repellents.

Repellent	Brand name	Formulation	Duration of protection[1] Mosquitoes, ticks	Cost (size)[2]
DEET	Cutter Sensations Repel Scented Family Off Deep Woods VIII Sawyer Ultra 30 Liposome Controlled release Ultrathon[4]	7% pump spray 15% aerosol spray 25% aerosol spray 30% lotion 34% lotion	1–3 h[3,] 6 h[3] 5–8 h[3], 8.5 h[3] 8 h[3], 5 h[3] 11 h, N.A. 12 h, N.A.	$6.40 (7.5 oz) 6.86 (6.5 oz) 7.38 (6 oz) 8.49 (4 oz) 6.99 (2 oz)
Picaridin	Cutter Advanced Avon Skin So Soft Bug Guard Plus Picaridin Natrapel 8 h	5.75% wipes 10% aerosol spray 20% pump spray	8 h, 5 h 8 h, 12 h 8 h, 8 h	5.99 (18 wipes) 6.99 (4 oz) 5.59 (3.4 oz)
IR3535	Avon Skin So Soft Bug Guard Plus IR3535[5] Coleman Skin Smart	7.5% lotion 20% pump spray	2 h, 2 h 8 h, 8 h	10.89 (4 oz) 4.99 (4 oz)
Oil of lemon eucalyptus	Coleman Botanicals Repel Lemon Eucalyptus	30% pump spray 30% pump spray	6 h, N.A. 7–8 h[3], 7 h[3]	8.94 (4 oz) 4.99 (4 oz)
Permethrin	Sawyer Premium Permethrin Clothing Repel Permethrin Clothing and Gear	0.5% pump spray 0.5% aerosol spray	– – – –	14.99 (24 oz) 7.42 (6.5 oz)

N.A. = not available

[1.] For repellents applied to exposed skin, according to protection times approved by the EPA for product labels. Available at: www.epa.gov/insect-repellents/find-insect-repellent-right-you. Accessed 23 June 2016. Duration of protection may vary depending upon ambient temperature, activity level, amount of perspiration, exposure to water, and other factors.

[2.] Cost at amazon.com (23 June 2016).

[3.] Duration of protection against *Aedes* and *Culex* mosquitoes and deer ticks according to the results of laboratory tests performed by Consumer Reports. Available at: www.consumerreports.org/cro/health/beauty-care/insect-repellent-ratings-overview.htm. Accessed 23 June, 2016.

[4.] Long-acting polymer-based formulation developed for the US military.

[5.] Contains IR3535 combined with sunscreen; products that contain both and insect repellent and a sunscreen are not recommended because the sunscreen may need to be reapplied more often and in greater amounts than the repellent.

Source: July 4, 2016. *The Medical Letter on Drugs and Therapeutics* **58** (1498), 83. Published by The Medical Letter, Inc. (a nonprofit organisation). New Rochelle, NY.

Appendix 4. Likely human allergens in botanical repellents

Active ingredient[1]	Human allergens[1]
Castor oil	None
Cedar oil	None
Citronella oil	Geraniol[2], citronellol[2], limonene[2], linalool[2], farnesol[2]
Clove oil	Eugenol[2]
Oil of lemon eucalyptus (PMD 65%)	Citronellol[2], pinene, caryophyllene
Geraniol oil	Geraniol[2], citronellol[2], linalool[2], limonene[2]
Lemongrass oil	Geraniol[2], citral[2], citronellol[2], limonene[2]
Peppermint oil menthol	Caryophyllene, limonene[2], pinene
Rosemary oil	Pinene, camphor, caryophyllene, limonene[2], terpineol, linalool[2]
Soybean oil	None

[1.] In Europe, cosmetics applied to the skin must disclose these ingredients when the concentration exceeds 0.001% European Commission Scientific Committee on Consumer Safety (SCCS) 2003.

[2.] Known human allergens identified by the European Commission Scientific Committee on Consumer Safety (SCCS) 2011.

Adapted from Shannon E. *et al.*, *EWG's Guide to Better Bug Repellents*, July 2013, Environmental Working Group, 1436 U Street NW, Suite 100, Washington, DC 20009.

Appendix 5. Personal health summary example

Your full name

Health summary

(Date)

1. Precautions: Note any serious medical condition (high blood pressure, diabetes, asthma, etc.)

Precaution	Description	Reaction
Drug allergies		
Other allergies	i.e. latex, cedar, oak and grass	
Surgical	pacemaker, prosthetic joint, organ transplant recipient, etc.	

2. Current routine medications:
   A. Prescription:

Name	Dose	Route	Frequency	Provider	Condition

   B. Over the counter:

Name	Dose	Route	Frequency	Provider	Condition

3. Current as needed medications:
   A. Prescription:

Name	Dose	Route	Frequency	Provider	Condition

B. Over the counter:

Name	Dose	Route	Frequency	Provider	Condition

4. Baseline Laboratory work: give the date of laboratory results
Comprehensive metabolic panel: complete blood count

Glucose _____   WBC _____

BUN _____   RBC _____

Creat. _____   HGB _____

GFR _____   HCT _____

Na^+ _____   MCV _____

K^+ _____   MCH _____

Cl^- _____   MCHC _____

CO_2 _____   RDW _____

Ph _____   PLT _____

Ca^{2+} _____   PSA _____

Albumin _____

Name	Relationship	Medical History

6. Past surgical history:

Date	Facility	Reason	Treatment

7. Major medical procedures

Date	Facility	Reason	Treatment

8. Hospitalisations for Past 10 Years:

Date	Facility	Reason	Treatment

9. Contacts: Next of kin, medical providers, travel agent.

Name	Relationship	Contact Informatio

Appendix 6. Medically important and venomous invertebrates of the world

Following the scientific name is the common name, if known, and the geographic distribution.

Sponges (Phylum Porifera)
    Family Mycalidae
      Neofibularia nolitangere (touch-me-not sponge): temperate to tropical oceans worldwide
    Family Tedaniidae
      Tedania ignis (fire sponge): temperate to tropical oceans worldwide

Coelenterates (Phylum Cnidaria: jelly fish, sea anemones, sea fans, coral)
    Family Cyanidae
      Cyanea capillata (lion's mane jellyfish): North Atlantic Ocean, Artic Sea
    Family Pelagidae
      Chrysaora achlyos (black sea nettle): Pacific Ocean; Baja California, southern California
      C. fuscescens (west coast sea nettle): Pacific Ocean; British Columbia, Mexico
      C. quinquecirrha (east coast sea nettle): Atlantic Ocean, Gulf of Mexico, Black Sea
      Pelagia noctiluca (pink jellyfish): oceans worldwide, Mediterranean Sea
    Family Ulmaridae
      Aurelia aurita (moon jellyfish): Gulf of Mexico, Atlantic Ocean
    Family Plumularidae
      Aglaophenia cupressina (cypress sea fern): Indo-West pacific
      Lytocarpus philippinus (white-stinging sea fern): Indo-West pacific
    Family Physaliidae
      Physalia physalis (Portuguese man o'war): subtropical oceans worldwide
    Family Chirodropidae
      Chiropsalmus quadrigatus (box jellyfish): Indo-Pacific Ocean
      Chironex fleckeri (box jellyfish, sea wasp): Indo-Pacific Ocean
    Family Carybdeidae
      Carybdea alata (box jellyfish): Southern Pacific Ocean
      C. rastonii (box jellyfish): Southern Pacific Ocean
      Carukia barnesi (box jellyfish): Southern Pacific Ocean
      Keesingia gigas (box jellyfish): Indian Ocean off Western Australia

Malo bella (box jellyfish): Indian Ocean off Western Australia

M. filipina (box jellyfish): Southern Pacific Ocean

M. kingi (box jellyfish): Southern Pacific Ocean

M. maximus (box jellyfish): Southern Pacific Ocean

Family Linuchidae

Linuche unguiculata (thimble jellyfish): Atlantic Ocean (Caribbean)

Family Discosomatidae

Amplexidiscus fenestrafer (balloon corallimorph anemone): western Pacific

Lytocarpus philippinus (white-stinging sea fern): Indo-West Pacific

Family Milleporidae

Millepora alcicornis (fire coral): warm oceans worldwide

M. complanata (fire coral): warm oceans worldwide

Millepora spp. (fire corals): warm oceans worldwide

Bryozoans (Phylum Bryozoa)

Family Alcyonidiidae

Alcyonidium gelatinosum: North Atlantic Ocean

Echinoderms (Phylum Echinodermata: sea stars and brittle stars)

Family Acanthasteridae

Acanthaster planci (crown of thorns starfish): coral reefs worldwide

Family Echinasteridae

Plectaster decanus (Mosaic sea star): Indo-Pacific Ocean

Family Ophiodermatidae

Ophiomastix annufosa (chain-link brittle star): Indo-Pacific Ocean

Annelid worms (Phylum Annelida: leeches, polychaetes)

Leeches

Family Glossophoniidae

Haementeria depressa: South America

H. ghilianii (Amazon leech): South America

H. officinalis: Central America

Family Haemadipsidae (land leeches)

Haemadipsa spp. (land leech): Eastern and South-eastern Asia

Family Hirudinidae

Hirudo medicinalis (European medicinal leech): Europe

H. nipponia: Asia

Hirudinaria manillensis (Asian medicinal leech): southern Asia, Philippines

Macrobdella decora: North America

Polychaetes

Family Amphinomidae

Eurythoë complanata (bristle worm): southern Pacific Ocean

Hermodice carrunculata (Caribbean fire worm): Caribbean

Family Goniadidae

Glycera spp. (blood worm): temperate and tropical oceans worldwide

Family Eunicidae
Eunice aphroditois: warm ocean water worldwide
Family Ohuphidae
Onuphis teres (beach worm): Australia
Nudibranchs and sea slugs
Family Aeolidiidae
Aeolidia spp.: oceans worldwide
Family Glaucidae
Glaucus spp.: oceans worldwide
Family Facelinidae
Hermissenda spp.: Pacific Ocean
Family Creseidae
Creseis acicula (sea slug): warm seas worldwide
Molluscs (Phylum Mollusca: bivalves, squid, octopus snails, cones)
Family Octopodidae
Hapalochlaena lunulata (greater blue-ringed octopus): tropical and subtropical seas
of Austral-Asia
H. maculosa (southern blue-ringed octopus): southern coast of Australia
H. fasciata (blue-lined octopus): coastal Australia, New South Wales to Queensland
Family Conidae
Conus aulicus (cone shell): southern Pacific Ocean
C. catus (cone shell): Southern Pacific Ocean
C. geographus (geography cone): southern Pacific Ocean
C. imperialis (cone shell): southern Pacific Ocean
C. lividus (cone shell): southern Pacific Ocean
C. marmoreus (cone shell): southern Pacific Ocean
C. nanus (cone shell): southern Pacific Ocean
C. obscurus (cone shell): southern Pacific Ocean
C. omaria (cone shell): southern Pacific Ocean
C. striatus (cone shell): southern Pacific Ocean
C. textile (cone shell): southern Pacific Ocean
C. tulipa (cone shell): southern Pacific Ocean
Family Bulimidae
Oncomelania spp.: Asia and Asian Pacific islands
Family Planorbidae
Biomphalaria (*Australorbis*) spp.: Africa, Middle East, Madagascar, Central and
South America, south-western United States, certain Caribbean islands
Bulinus spp.: Africa, Middle East, Madagascar, Mauritius, India
Arthropods (Phylum Arthropoda)
Arachnids (Class Arachnida)
Spiders (Class Arachnida, Order Araneae)
Banana spiders (Family Ctenidae)
Phoneutria fera: Central and South America

P. ochracea: Central and South America
Black widows (Family Theridiidae)
Latrodectus antheratus: Paraguay, Argentina
L. apicalis: Galapagos Islands
L. atritus: New Zealand
L. bishopi (red widow): south-eastern United States
L. cinctus: Cape Verde Island, South Africa
L. corallinus: Argentina
L. curacaviensis (Brazilian black widow): Lesser Antilles, Americas
L. dahli: southern Europe, northern Africa, Middle East
L. diaguita: Argentina
L. elegans: India, Myanmar, China, Japan
L. erythromelas: Sri Lanka
L. geometricus (brown widow): southern United States, South Africa, Japan, South-East Asia
L. hasselti (redback): South-East Asia, Australia, New Zealand, Japan, Marianas, Philippines, Timor, many other regional islands, New Guinea
L. hesperus (western black widow): western North America, Israel, Singapore
L. hystrix: southern Europe, northern Africa, Middle East
L. indistinctus: Namibia, South Africa
L. karrooensis: South Africa
L. katipo: New Zealand
L. lilianae: Spain
L. mactans (southern black widow): southern North America, South Africa
L. menavodi: Madagascar, Comoro Islands
L. mirabilis: Argentina
L. obscurior: Cape Verde Islands, Madagascar
L. pallidus: Cape Verde Islands, Libya, Europe, Russia, Iran, Africa, Middle East, Turkey
L. quartus: Argentina
L. renivulvatus: Africa, Saudi Arabia, Yemen, Iraq
L. reviensis: Israel
L. rhodesiensis: South Africa
L. tadzhicus: Tajikistan
L. tredecimguttatus: Mediterranean, Saudi Arabia, Ethiopia, South Africa, central Asia
L. variegates: Chile, Argentina
L. variolus (northern black widow): northern North America
Brown recluses (Family Siciariidae)
Loxosceles laeta: South America
L. parrami: South Africa
L. reclusa (brown recluse): North America east of the Rocky Mountains
L. rufescens: Europe, Middle East, Australia, Madagascar, Japan, North America

L. rufipes: Guatemala, Panama, Colombia

L. spinulosa (savanna violin spider): southern Africa

L. speluncarum: South Africa

L. unicolor: South America

Funnel web spiders (Family Hexathelidae)

Atrax robustus (Sydney funnel web spider): New South Wales, Australia

Hadronyche cerberea: New South Wales, Australia

H. formidabilis: Queensland and New South Wales, Australia

H. infensa: Queensland and New South Wales, Australia

H. modesta: Victoria, Australia

H. versuta: New South Wales, Australia

Mouse spiders (Family Actinopodidae)

Missulena tussulena: Chile

Missulena spp. (16 species): Australia

Megalomorph and baboon spiders (Family Theraphosidae)

Harpactirella lightfooti: South Africa

Pelinobius muticus: eastern Africa (Kenya, Tanzania)

Six-eyed sand/crab spiders (Family Sicariidae)

Sicarius spp.: Zimbabwe, southern Africa, Central and South America, Galapagos Islands.

Wandering spiders (Family Heteropodidae)

Palystes superciliosus: South Africa

White-tailed spider (Family Lamponidae)

Lampona cylindrata: Australia

L. murina: Australia

Yellow sac spiders (Family Miturgidae)

Cheiracanthium brevicalcaratum: Western Australia

C. fulcatum: South Africa

C. inclusum: North, Central and South America, West Indies, United States, south-western Canada

C. japonicum: Japan

C. mildei: Mediterranean, northern Europe, eastern Canada, north-eastern United States

C. mordax: eastern and western Australia, central and south-western Pacific, United States (including Hawaii)

C. punctorium: Europe

Other spiders of potential concern

Argiope spp. (Family Araneidae, garden spiders): worldwide

Phidippus spp. (Family Salticidae, jumping spiders): worldwide

Lycosa raptoria (Family Lycosidae, wolf spider): South America

Scorpions (Order Scorpiones, specific states in parentheses where appropriate)

Family Buthidae

Androctonus amoreuxi: North Africa

A. australis (fat-tailed scorpion): Middle East, North Africa

A. bicolor (black fat-tailed scorpion): North Africa

A. crassicauda: Middle East

Buthus occitanus (yellow thick-tailed scorpion): Mediterranean, North Africa

Centruroides elegans: Mexico (Jalisco)

C. exilicauda (Baja California bark scorpion): Mexico (Baja California)

C. limpidus: western Mexico

C. noxius: Mexico (Nayarit)

C. sculpturatus: United States (Arizona, California, Utah), Mexico (Sonora)

C. suffusus: Mexico (Durango)

Compsobuthus acuticarinatus: Egypt

Hottentota saulcyi: Iran

H. tamulus (red scorpion): India

Leiurus quinquestriatus (yellow scorpion or death stalker): northern Africa, Middle East

Mesobuthus eupeus: Iran

Odontobuthus doriae: Iran

Parabuthus granulatus: southern Africa

P. transvaalicus (fat-tailed scorpion): southern Africa

Tityus bahiensis: Brazil

T. serrulatus (Brazilian yellow scorpion): Brazil

T. trinitatis: Trinidad

Family Hemiscorpiidae

Hemiscorpius lepturus: Iran

Family Scorpionidae

Opistophthalmus glabrifrons (yellow creeping leg scorpion): southern Africa

Mites (Subclass Acari)

Chigger mites (Family Trombiculidae)

Leptotrombidium spp.: Asia, Indonesia, Australia, Philippines

Eutrombicula alfreddugesi (chigger): southern North America

Trombicula autumnalis: Europe

T. splendens: southern North America

Scabies mites (Family Sarcoptidae)

Sarcoptes scabei (scabies): worldwide

Dust mites (Family Pyroglyphidae)

Dermatophagoides farinae (dust mite): worldwide

D. pteronyssinus (European house dust mite): worldwide

Chicken mite (Family Cheyletiellidae)

Dermanyssus gallinae (chicken mite): worldwide

Bird and fowl mites (Family Macronyssidae)
Ornithonyssus bacoti (tropical rat mite): worldwide
O. bursa (tropical fowl mite): worldwide
O. sylviarum (northern fowl mite): temperate areas worldwide
Rabbit fur mite (Family Cheyletiellidae)
Cheyletiella parasitivorax
Spiny rat mite (Family Laelapidae)
Laelaps echidnina (spiny rat mite): worldwide
House mouse mite (Family Dermanyssidae)
Liponyssoides sanquineus (house mouse mite): worldwide
Oak leaf itch mite (Family Pyemotidae)
Pymotes herfsi (oak leaf itch mite): Australia, Europe, India, United States
Straw itch mite (Family Pyemotidae)
Pyemotes tritici (straw itch mite): temperate areas worldwide
Grain and flour mite (Family Acaridae)
Acarus siro (grain or flour mite): worldwide

Ticks (Subclass Acari)

Hard ticks (Family Ixodidae)
Amblyomma americanum (lone star tick): central and eastern United States, Mexico
A. cajennense (Cayenne tick): southern United States, Mexico, Central and South America
A. hebraeum (South African bont tick): central and southern Africa
A. lepidum: east Africa
A. maculatum (gulf coast tick): North America, Caribbean and Gulf of Mexico
A. variegatum (tropical bont tick): Africa and Caribbean
Boophilus spp. (some consider this a subgenus of *Rhipicephalus*) (cattle ticks): worldwide
Dermacentor albipictus (winter tick): North America
D. andersoni (Rocky Mountain wood tick): western United States, Canada
D. auratus: Asia
D. marginatus: Europe, western Asia
D. nuttalli: Eastern Europe, northern Asia
D. parumapertus (rabbit Dermacentor): United States
D. occidentalis (Pacific Coast tick): western United States (California), Mexico (Sonora)
D. reticulatus (ornate cow tick): Europe, eastern Asia from Atlantic coast to Kazakhstan, central Africa.
D. silvarum: Europe, northern Asia
D. taiwanensis: Japan, Taiwan
D. variabilis (American dog tick): United States, Mexico
Haemaphysalis bispinosa: India
H. concinna: Europe, Asia

H. inermis (winter tick): Europe

H. intermedia: India

H. leachi (yellow dog tick): Africa, Asia, (eastern Australia?)

H. leporispalustris (rabbit tick or grouse tick): Americas (Alaska to Argentina)

H. longicornis: Russia

H. punctata: Europe

H. spinigera: India, South-East Asia, Indonesia

H. wellingtoni: India

Hyalomma aegptium (Tortoise tick): southern Europe

H. asiaticum (Asiatic Hyalomma): Asia

H. anatolicum anatolicum: Europe, Asia, India, Middle East, Africa

H. impeltatum: Mediterranean, north Africa, Sudan, Eritrea, Somalia, northern Kenya, northern Tanzania, Chad, the Middle East

H. isaaci: Asia

H. marginatum: Africa, Asia, Europe, India

H. rufipes: Nigeria

H. truncatum (small smooth bont-legged tick): southern Africa

Ixodes acutitarsus: Himalayas to southern Japan

I. ceylonensis: India

I. granulatus: Japan through South-East Asia and westward to India and China

I. holocyclus (Australian paralysis tick): Australia, Papua New Guinea

I. pacificus (western black-legged tick): western United States, Mexico

I. pavlovskyi: central and eastern Europe, northern Asia

I. persulcatus (Taiga tick): central and eastern Europe, northern Asia

I. petauristae: India

I. redikorzevi: Israel

I. ricinus (European castor bean tick): Europe, northern Africa, northern Asia

I. rubicundus: South Africa

I. scapularis (black-legged tick): central and eastern United States, Mexico

Nosomma monstrosum: India, Bangladesh, South-East Asia, Nepal, Sri Lanka

Rhipicephalus appendiculatus (brown ear tick): central and southern Africa

R. sanguineus (brown dog tick): worldwide

R. pumilio: Iberian Peninsula, Morocco, north-west Africa, southern France, Italy, Sicily, Ustica, Russia

R. pulchellus: East Africa

R. turanicus: tropical Africa, southern Europe, Arabia, Asia

R. bursa (brown ear tick): Mediterranean, Switzerland, Bulgaria, Romania

Soft ticks (Family Argasidae)

Argas reflexus (pigeon tick): African

A. brumpti: Old World

A. persicus: western United States

A. vespertilionis (round bat argasid): Old World, most recently Iran

Ornithodoros chiropterphila: India

O. coniceps: central Asia

O. coriaceus (Parjaroella tick): western United States, Mexico

O. erraticus: North Africa, Near East, Spain, Portugal, southern Russia

O. erraticus sonrai: Africa, Far East, Central Asia

O. hermsi: western United States and Canada

O. marocanus: south-western Europe, north-western Africa

O. moubata (eyeless Tampan): Africa

O. parkeri: western United States

O. rudis: Central and South America

O. talaje: southern and western United States, Mexico, Central and South America

O. tholozani: Uzbekistan, Kashmir to Cyprus, Tripoli

O. turicata (relapsing fever tick): Central, southern and western United States, Mexico

O. verrucosus: Caucasus

Otobius megnini (Persian fowl-tick): western United States, Mexico, western Canada, western South America, Galapagos, Cuba, Hawaii, India, Madagascar, south-eastern Africa

Camel Spiders (Order Solifugae, 'those who flee from the sun'): Africa, Middle East, Mexico and south-western United States

Centipedes (Class Chilopoda)

Family Scolopendridae

Scolopendra spp.: subtropical and tropical areas worldwide

Family Otostigmidae

Otostigmus spp.: tropical areas worldwide

Millipedes (Class Diplopoda)

Family Rhinocrichidae

Rhinocricus lethifer: Haiti

R. latespargor: Haiti

Family Spirobolidae

Julus spp.: Indonesia

Spirobolus spp.: Tanzania

Tylobolus spp.: California

Family Spirostreptidae

Orthoporus spp.: Mexico, Central and South America

Polyceroconas spp.: Papua New Guinea

Spirostreptus spp.: Indonesia

Crustaceans (Class Crustacea)

Order Decapoda (shrimps, lobsters, crabs)

Insects (Class Insecta)

Lice (Order Phthiraptera)

Family Pediculidae
Pediculus humanus capitis (head louse): worldwide
P. humanus humanus (body louse): worldwide
Family Pthiridae
Pthirus pubis (pubic louse): worldwide
Cockroaches (Order Blattodea)
Family Blattidae
Blatta orientalis (oriental cockroach): worldwide
Periplaneta americana (American cockroach): Americas
P. fuliginosa (smoky brown cockroach): Americas
Family Blattellidae
Blatella asahinae (Asian cockroach): worldwide
B. germanica (German cockroach): worldwide
Supella longipalpa (brown-banded cockroach): worldwide
True bugs (Order Hemiptera)
Bed bugs (Family Cimicidae)
Cimex hemipterus: tropical areas worldwide
C. lectularius (bed bug): worldwide
Assassin and kissing bugs (Family Reduviidae)
Arilus cristatus (wheel bug): Americas
Arilus spp.: Americas
Pristhesancus plagipennis: Queensland, Australia
Reduvius personatus (Masked hunter): Americas
Panstrongylus spp.: Central and South America
Rhodnius spp.: Central and South America
Triatoma spp.: south-western United States, Mexico, Central and South America
Water boatmen (Family Corixidae): worldwide
Giant water bugs (Family Belastomatidae): worldwide
Backswimmers (Family Notonectidae): worldwide
Ants, bees, wasps, hornets (Order Hymenoptera)
Ants (Family Formicidae)
Myrmecia pilosa (jumper ant): Australia
M. gulosa (bull-dog ant): Australia (including Tasmania), New Caledonia
M. pyriformis (bull-dog ant): Australia (including Tasmania), New Caledonia
Pachycondyla sennaarensis (samsun ant): Middle East
Paraponera clavata (bullet ant): Central and South America
Pogonomyrmex spp. (harvester ants): south-western United States, Mexico, Central and South America
Solenopsis invicta (red-imported fire ant): southern United States, Mexico, Central and South America
S. richteri (black-imported fire ant): southern United States, Mexico, Central and South America

Tetramorium caespitum (pavement ant): Europe, North America

Odontomachus (trap-jaw ants): Central and South America, Asia, Australia, Africa and the south-eastern United States

Wasps, hornets and yellow jackets (Family Vespidae): worldwide

Dolichovespula maculata (bald-faced hornet): North America

Vespa crabo (European hornet, sand hornet, brown hornet, German hornet): Europe, North America

V. mandarinia (Asian giant hornet): eastern and South-East Asia

V. orientalis (Oriental hornet): south-west Asia, China, North-east Africa, Madagascar, Middle East, Mexico

V. velutina: South-East Asia, South Korea, Japan, Europe

Vespula spp. (yellow jackets): Northern Hemisphere

Velvet Ants (Family Mutillidae): worldwide

Honey and bumble bees (Family Apidae)

Apis mellifera (honey bee): worldwide

Bombus spp. (bumble bees): worldwide

Xylocopa spp. (carpenter bees): worldwide

Butterflies and moths (Order Lepidoptera)

Family Limacodidae

Calcarifera ordinata: Australia

Doratifera vulnerans (spitfire): Australia

Euclea delphinii (spiny oak slug): North America

Isa textula (crowned slug): North America

Latoia consocia: Asia

Parasa indetermina (stinging rose caterpillar): North America

Parasa spp.: Americas, Japan, Asia, New Zealand

Phobetron pithecium (hag moth/monkey slug): Americas

Sabine stimulea (saddle-back caterpillar): North America

Family Lymantriidae

Euproctis spp. (cup moth caterpillar): Australia

Euproctis chrysorrhoea (browntail moth): Europe

E. flava (oriental tussock moth): Japan

E. pseudoconspersa (tea tussock moth): Asia, Japan, Australia

E. silimis (yellowtail moth): Japan

Lymantria dispar (gypsy moth): North America

Family Megalopygidae

Megalopyge opercularis (brown puss caterpillar or flannel moth): North America

Megalopyge spp.: Americas

Family Saturniidae

Adeloneivia spp.: Central and South America

Automeris io (io moth): Americas

Automeris spp.: Americas

Cerodirphia spp.: South America

Dirphia spp.: Central and South America
D. panamensis: Central America
Hemileuca spp. (other): south-western United States, Mexico
H. maia (buck moth): eastern United States
Hylesia alinda: Cozemel, Mexico, Venezuela, Peru
H. ebalus: South America
H. iola: South America
H. lineata: South America
H. metabus: Venezuela
H. urticans: South America
Hyperchiria spp.: Mexico to South America
Leucanella lama: Central and South America
Lonomia achelous: northern South America
L. obliqua: Brazil
Molippa basina: South America
Molippa spp.: Mexico to South America
Family Thaumetopoeidae
Anaphae venata: Africa
A. panda: Africa
Beetles (Order Coleoptera)
Family Meloidae
Epicauta spp.: Americas
E. vittata: North America
Lytta spp.: Australia, Eurasia, North America
L. vesicatoria: Europe
Mylabris quadripunctata: Middle East
Mylabris spp.: Asia, Australia, Europe, Middle East, New Zealand
Family Staphylinidae
Paederus fuscipes (rove beetle): Asia
Family Dermestidae
Trogaderma granarium (Khapra beetle): worldwide
Flies (Order Diptera)
Black flies (Family Simuliidae)
Simulium spp.: worldwide
Cnephia spp.: temperate areas worldwide
Blow and bottle flies (Family Calliphoridae)
Auchmeromyia senegalensis (Congo floor maggot): Africa
Chrysomya bezziana (Old World screwworm): Africa, southern Asia
C. chloropyga (sheep maggot, smooth maggot): South Africa
C. megacephala (Old World latrine fly): Indo-Australian, Afrotropical
Chrysomya spp.: Asia, Australia, Indonesia
Cochliomyia hominivorax (New World screwworm): Central and South America
Cordylobia anthropophaga (tumbu fly): central and tropical Africa

C. rodhaini (Lund's fly): central and tropical Africa
Cosmina bicolor: Asia
Lucilia (= *Phaenicia*) *cuprina* (sheep blow fly): Asia, Australia
L. sericata (green-bottle fly): worldwide
Phormia regina (black blow fly): worldwide
Eye gnats (Family Chloropidae)
Hippelates spp.: worldwide
Liohippelates spp.: worldwide
  Siphunculina spp.: worldwide
  Flesh flies (Family Sarcophagidae)
  Wohlfahtria magnifica (spotted flesh fly): Europe (Bosnia and Herzegovina, Bulgaria, Croatia, France, Hungary, Italy, Hungary, Macedonia, Montenegro, Romania, Serbia, Slovakia, Slovenia, Spain), southern and Asiatic Russia, Middle East (Egypt, Israel, Iran, Turkey), North Africa, and Asia (China, Mongolia).
Horse and deer flies (Family Tabanidae)
Chrysops spp. (deer flies): temperate and tropical areas worldwide
Tabanus spp. (horse flies): temperate and tropical areas worldwide
Human bot fly (Family Cuterebridae)
Dermatobia hominis (human bot fly): Mexico, Central and South America
Humpbacked flies (Family Phoridae)
Megaselia scalaris: worldwide
Midges, no-see-ums, punkies (Family Ceratopogonidae)
Culicoides spp.: worldwide
Leptoconops spp.: worldwide
Mosquitoes (Family Culicidae)
Aedes: worldwide
Anopheles: worldwide see IAMAT World Malaria Risk Chart (https://www.iamat.org).
Culex spp.: worldwide
Haemagogus spp.: Central and South America
Mansonia spp.: worldwide
Ochlerotatus spp.: worldwide
Psorophora spp.: worldwide
Fanniid flies (Family Fanniidae)
Fannina canicularis: Worldwide
Fannia scalaris: Asia
Muscoid flies (Family Muscidae)
Haematobia exigua: Asia
Lispe orientalis: Asia
Musca biseta: Africa
M. conducens: Asia
M. crassirostris: Asia
M. domestica (house fly): worldwide

M. fasciata: Asia

M. lucens: Asia

M. sorbens (dog dung fly, bazaar fly): Afrotropical and oriental regions, Australia; introduced elsewhere, including Hawaii

M. ventrosa: Asia

M. vetustissima (Australian bush fly): Australia

Muscina stabulans: Asia

Stomoxys calcitrans (stable fly): worldwide

S. niger: Afrotropical region

S. sitiens: Afrotropical and oriental regions

S. uruma: Asia

Sand flies (Family Psychodidae)

Lutzomyia spp.: south-western United Status, Mexico, Central, South America

Phlebotomus spp.: Africa, Asia, Bangladesh, Europe, India, Middle East, Pakistan

Tsetse flies (Family Glossinidae)

Glossina spp. (tsetse fly): Africa

Warble and bot flies (Family Oestridae)

Oestris ovis (sheep bot fly): Middle East, Cyprus

Fleas (Order Siphonaptera)

Family Ceratophyllidae

Ceratophyllus gallinae (European chicken flea): temperate areas worldwide

C. niger (western chicken flea): western North America

Dasypsyllus gallinulae (bird flea): worldwide

Diamanus montanus (ground squirrel flea): western North America

Leptopsylla segnis (European mouse flea): worldwide

Nosopsyllus fasciatus (northern rat flea): North America and Europe

Orchopeas howardii (squirrel flea): North America

Family Hystrichopsyllidae

Neopsylla setosa (rodent flea): Eurasia

Family Pulicidae

Cediopsylla simplex (rabbit flea): North America

Ctenocephalides canis (dog flea): worldwide

C. felis (cat flea): worldwide

Echidnophaga gallinacea (sticktight flea): worldwide

Hoplopsyllus anomalus (rodent flea): North America

Pulex irritans (human flea): temperate and tropical areas worldwide

Tunga penetrans (jigger, chigoe, chique, chigger or sand flea): tropical and subtropical Africa, India, the Caribbean, North and South America

Xenospylla astia: India, Myanmar

X. brasiliensis: South America, India, Africa (Uganda, Kenya, Nigeria)

X. cheopsis (Oriental rat flea): worldwide

X. vexabilis (Australian rat flea): Australia

Appendix 7. Common vector-borne diseases, their vectors, distribution, symptoms, vaccines and chemoprophylaxis[1]

Vector group	Disease	Pathogen (type and species)	Major vector(s)	Distribution	Incubation period and initial symptoms	V, C, L[2]
Mosquitoes family Culicidae	Malaria	Protozoans Plasmodium falciparum P. malarie P. ovale P. vivax	Anopheles spp.	Worldwide: tropical and temperate areas	10–21 days: fever and chill cycles, malaise, myalgia, headache, anorexia, fatigue	C
	Filariasis	Nematodes Wucheria brancofti	Aedes aegypti Ae. vigilax Ae. niveus Ae. poicilius Ae. s. pseudoscutellaris Ae. scapularis Ae. s. scutellaris Anopheles gambiae An. funestus Culex annulirostris C. quinquefasciatus	Africa, Asia, Central and South America, Pacific Islands	1–22 months: localised pain, tenderness, swelling, numbness, weakness, mild fever, erythema, malaise, headache, insomnia, nausea	N/A
		Brugia malayi	Anopheles barbirostris Aedes togoi Mansonia annulata M. annulifera M. uniformis M. indiana M. longipalpis M. dives M. bonneae	Asia, India, Sri Lanka	Same as above	N/A

Vector group	Disease	Pathogen (type and species)	Major vector(s)	Distribution	Incubation period and initial symptoms	V, C, L²
Mosquitoes family Culicidae (Continued)		*Brugia timori*	*Anopheles barbirosis*	Indonesia	Same as above	N/A
	Dengue (D), D haemorrhagic fever (DHF), D shock syndrome (DSS)	Virus family Flaviviridae	*Aedes aegypti* *Ae. albopictus* *Ae. scutellaris* *Ae. niveus* *Ae. furcifer-taylori*	Urban and rural Old and New World tropical areas (>61 countries), from 40°North to 40°South	3–14 days: high, biphasic fever; fatigue; rash; retro-orbital pain; body aches; DHF/DSS much more severe	V/L
	Yellow fever	Virus family Flaviviridae	**Urban** *Aedes aegypti* *Ae. albopictus*	Central and South America, tropical Africa	3–5 days: high phasic fever; chills; fatigue; headache; back pain; nausea; vomiting	V
			Sylvatic *Aedes aegypti* *Ae. albopictus* *Ae. africanus* *Ae. simpsoni* *Ae. furcifer-taylori* *Haemagogus* spp. *Sabethes* spp.	Central and South America, tropical Africa	Same as above	V
	West Nile fever	Virus family Flaviviridae	*Aedes* spp. *Culex* spp.	Africa, Europe, Russia, North America, Pakistan, Middle East, Indian subcontinent	3–5 days: fever, malaise, fatigue, encephalitis, rash	N/A
	Japanese encephalitis	Virus family Flaviviridae	*Culex* spp.	Asia, Australia, Pacific Islands	5–15 days: fever, malaise, fatigue, encephalitis	V/L
	Murray Valley encephalitis	Virus family Flaviviridae	*Aedes* spp. *Culex* spp.	Australia, New Guinea	5–15 days: malaise, fatigue, encephalitis	N/A
	St. Louis encephalitis	Virus family Flaviviridae	*Culex* spp.	North through South America	5–15 days: malaise, fatigue, hepatitis, encephalitis	N/A
	Rocio fever	Virus family Flaviviridae	*Aedes* spp. *Psorophora* spp.	Brazil	5–15 days: malaise, fatigue, encephalitis	N/A
	Ross River fever	Virus family Togaviridae	*Aedes* spp. *Anopheles* spp. *Culex* spp. *Mansonia* spp.	Australia, South Pacific Islands	3–11 days: fever, malaise, arthralgia, rash	N/A

Vector group	Disease	Pathogen (type and species)	Major vector(s)	Distribution	Incubation period and initial symptoms	V, C, L[2]
Mosquitoes family Culicidae (Continued)	Eastern equine encephalitis	Virus family Togaviridae	*Culex* spp. *Culiseta* spp. *Ochlerotatus* spp.	North through South America	5–15 days: malaise, fatigue, encephalitis	V/L
	Western equine encephalitis	Virus family Togaviridae	*Culex* spp. *Ochlerotatus* spp.	western North America	5–15 days: fever, malaise, fatigue, encephalitis	V/L
	Venezuelan equine encephalitis	Virus family Togaviridae	*Aedes* spp. *Culex* spp. *Psorophora* spp.	North through South America	2–6 days: fever, malaise, fatigue, encephalitis	V/L
	Sinbis fever	Virus family Togaviridae	*Culex* spp.	Europe, Africa, Russia, Southeast Asia, Philippines, Australia	3–11 days: fever, rash malaise, arthralgia	N/A
	O'Nyong-nyong fever	Virus family Togaviridae	*Anopheles* spp.	Tropical Africa	3–11 days: fever, rash malaise, arthralgia	N/A
	Chikungunya fever	Virus family Togaviridae	*Aedes* spp. *Mansoni* spp.	Africa, Caribbean, Southeast Asia, Philippines, Mediterranean	3–11 days: fever, malaise, fatigue, arthralgia, rash (rarely haemorrhage)	N/A
	Barmah Forest virus	Virus family Togaviridae	*Aedes* spp. *Culex* spp. *Coquillettidia* spp.	Australia	3–11 days: fever, rash malaise, arthralgia	N/A
	Semliki Forest virus	Virus family Togaviridae	*Aedes* spp. *Anopheles* spp.	Sub-Saharan Africa	3–11 days: malaise, fatigue, encephalitis	N/A
	California encephalitis	Virus family Bunyaviridae	*Ochlerotatus* spp.	USA	5–15 days: malaise, fatigue, encephalitis	N/A
	Bunyamwera virus	Virus family Bunyaviridae	*Aedes* spp. *Culex* spp. *Mansonia* spp.	Eastern, central and western Africa	3–12 days: fever, rash, malaise, fatigue	N/A
	Cache Valley virus	Virus family Bunyaviridae	*Aedes* spp. *Anopheles* spp. *Psorophora* spp.	North America	1–3 days: fever, malaise, myalgia, fatigue, encephalitis	N/A

Vector group	Disease	Pathogen (type and species)	Major vector(s)	Distribution	Incubation period and initial symptoms	V, C, L[2]
Mosquitoes family Culicidae (Continued)	Jamestown Canyon virus	Virus family Bunyaviridae	*Aedes* spp. *Culiseta* spp.	North America	5–15 days: malaise, fatigue, encephalitis	N/A
	LaCrosse encephalitis	Virus family Bunyaviridae	*Ochlerotatus* spp.	North America	5–15 days: malaise, fatigue, encephalitis	N/A
	Snowshoe hare virus	Virus family Bunyaviridae	*Aedes* spp.	North America, China, Russia	5–15 days: malaise, fatigue, encephalitis	N/A
	Tahyna virus	Virus family Bunyaviridae	*Aedes* spp. *Culex* spp.	Central Europe, Asia, Africa	3–11 days: fever, malaise	N/A
	Trivittatus virus	Virus family Bunyaviridae	*Ochlerotatus* spp.	North America	3–11 days: fever, malaise	N/A
	Rift Valley fever	Virus family Bunyaviridae	*Aedes* spp. *Anopheles* spp. *Culex* spp.	Africa	3–12 days: fever, malaise, fatigue, encephalitis, retinitis, haemorrhage	N/A
	Bovine ephemeral fever	Virus family Rhabdoviridae	*Aedes* spp. *Culex* spp. *Culicoides* spp.	Africa, Asia, Australia, Middle East	3–11 days: fever, malaise	N/A
Sand flies family Psychodidae	**Cutaneous leishmaniasis:** Aleppo, Bagdad or Dehli Boil; Oriental sore, Espundia; Uta or Chicken Ulcer **Visceral leishmaniasis:** Kala-azar, Dum Dum Fever, or Ponos Mucocutaneous leishmaniasis	Protozoans *Leishmania aethiopic* *L. amazonensis* *L. braziliensis* *L. chagasi* *L. donovani* *L. guyanensis* *L. infantum* *L. major* *L. minor* *L. mexicana* *L. panamensis* *L. peruviana* *L. tropica*	**Old world** *Phlebotomus* spp. **New world** *Lutzomyia* spp.	**Old world** Pakistan, India, China, sub-Saharan Africa, Middle East, southern Russia, Mediterranean **New world** Central and South America (not Chile and Uruguay), USA (Texas)	**Cutaneous** = 7–56 days: persistent skin lesion, discomfort **Mucosal** = 1 month to years: mucosal lesion **Visceral** = 2–6 months: fever, chills, malaise, nosebleed, cough, gastrointestinal discomfort, weight loss	N/A

Vector group	Disease	Pathogen (type and species)	Major vector(s)	Distribution	Incubation period and initial symptoms	V, C, L[2]
Sand flies family Psychodidae (Continued)	Sand fly fever (phlebotomus, pappataci, changuinola, vesicular stomatitis or three-day fever)	Virus family Bunyaviridae	Phlebotomus spp. Lutzomyia spp.	Worldwide	3–6 days: fever, malaise, headache, fatigue, body pain, retro-orbital pain, photophobia, vomiting, nausea, myalgia, weakness	N/A
Biting midges family Ceratopogonidae	Bartonellosis	Bacteria Bartonella spp.	Lutzomyia spp.	South America, Andes mountains	16–22 days: fever, malaise, fatigue, headache, myalgia, arthralgia, anemia	N/A
	Mansonellosis	Nematode Mansonella ozzardi	Culicoides spp.	Western and central Africa, American tropics	Weeks to months: fever, malaise, headaches, fatigue, pruritus, arthralgia, papules, angioedema	N/A
	Nairobi sheep disease	Virus family Bunyaviridae	Culicoides spp.	Eastern Africa, Somalia, India	4–5 days: fever, malaise	N/A
	Oropouche virus	Virus family Bunyaviridae	Culicoides spp.	Panama, Peru, Amazon Basin	3–12 days: fever, malaise, encephalitis	N/A
	Rift Valley fever	Virus family Bunyaviridae	Culicoides spp.?	Africa	3–12 days: fever, malaise, fatigue, encephalitis, retinitis, haemorrhage	N/A
Blackflies family Simuliidae	Onchocerciasis (river blindness)	Nematode Onchocerca volvulus	Simulium callidum S. damnosum S. exiguum S. metallicum S. neavei	Sub-Saharan Africa, Yemen, South America, southern Mexico, Guatemala,	>1 year: malaise, skin nodules, myalgia	C/L
	Rift Valley fever	Virus family Bunyaviridae	Simulium spp.	Africa	3–12 days: fever, malaise, fatigue, encephalitis, retinitis, haemorrhage	N/A
	Mansonellosis	Nematode Mansonella ozzardi	Simulium spp.	West and central Africa, American tropics	Weeks to months: fever, malaise, headaches, fatigue, pruritus, arthralgia, papules, angioedema	N/A

Vector group	Disease	Pathogen (type and species)	Major vector(s)	Distribution	Incubation period and initial symptoms	V, C, L²
Tsetse flies family Glossinidae	African trypanosomiasis 'African sleeping sickness'	Protozoan *Trypanosoma brucei*	**Riverine form:** *Glossina fuscipes* *G. palpalis* *G. tachinoides* **Rhodesian form:** *G. morsitans* *G. pallidipes* *G. swynnertoni*	Sub-Saharan Africa (14°N–20°S), Zanzibar	5–20 days: fever, malaise, fatigue, lymphadenopathy	C/L
Deer flies family Tabanidae	Loiasis	Nematode *Loa loa*	*Chrysops dimidiatus* *C. silacea*	Tropical western and central Africa	Usually years: localised swelling, pain, pruritus, oedema	N/A
	Anthrax (cutaneous form)	Bacteria *Bacillus anthracis*	*Chrysops* spp.	Worldwide	1–7 days: itching skin lesion, oedema	V/L
	Tularemia (rabbit fever, Ohara's disease, deer-fly fever)	Bacteria *Francisella tularensis*	*Chrysops* spp.	Worldwide	1–14 days: usually itchy skin lesion, oedema, swollen glands	N/A
Eye gnats family Chloropidae	Conjunctivitis 'pinkeye'	Bacteria *Streptococcus pyogenes*	Various species	Worldwide	24–72 hours: reddening of eye with irritation, discharge, swollen eyelids	N/A
	Yaws	Bacteria *Treponema pallidum*	Various species	Africa, Asia, India, Latin America	2 weeks–3 months: skin lesions which spread	N/A
Myiasis producing flies family Calliphoridae	Human bot fly (torsalo, moyocuil, berne, mucha, mirunta, ura)	Fly maggots Myiasis (obligate)	*Dermatobia hominis*	South Pacific islands, Mexico, Central and South America	Rapid: swollen, painful, itchy, warble like sub-cutaneous skin lesion (cyst) lasting 5–12 weeks	N/A
Flesh flies family Sarcophagidae	New World primary screwworm	Fly maggots Myiasis (obligate)	*Cochliomyia hominivorax*	South America (not Chile), Cuba, Haiti, Jamaica, Dominican Republic, Trinidad, Tobago	Rapid: invades open wounds forming larval filled pocket lasting 5–7 days	N/A
	New World secondary screwworm	Fly maggots Myiasis (facultative)	*Cochliomyia macellaria*	USA, southern Canada, Mexico, American tropics	Same as above	N/A

Vector group	Disease	Pathogen (type and species)	Major vector(s)	Distribution	Incubation period and initial symptoms	V, C, L[2]
Flesh flies family Sarcophagidae (Continued)	Old World primary screwworm	Fly maggots Myiasis (obligate)	*Chrysomya bezziana*	Africa, India, Pacific Islands	Same as above	N/A
	Old World secondary screwworm	Fly maggots Myiasis (facultative)	*Chrysomya ruffacies*	Europe, Asia, Australia	Same as above	N/A
	Screwworm	Fly maggots Myiasis (obligate)	*Wohlfahrtia magnifica*	Palearctic region	Same as above	N/A
	Screwworm	Fly maggots Myiasis (obligate)	*Wohlfahrtia vigil*	Nearctic region	Same as above	N/A
	Screwworm	Fly maggots Myiasis (obligate)	*Cordylobia rodhaini*	Northern and central Africa	Same as above	N/A
	Congo floor maggot	Fly maggots Nocturnal blood suckers	*Auchmeromyia senegalensis*	Northern and central Africa	15–20 minutes; swollen, itchy, skin lesions	N/A
Kissing bugs family Reduviidae	American Trypanosomiasis (Chagas disease)	Protozoan *Trypanosoma cruzi*	*Rhodnius* spp. *Triatoma* spp. *Panstrongylus* spp.	Central and South America, USA (Texas)	Rapid: you find hard, swollen, violet-hued skin lesion or swollen eye upon awakening	N/A
Sucking lice family Pediculidae	Louse-borne epidemic typhus	Bacteria *Rickettsia prowakekii*	*Pediculus humanus humanus*	Colder regions worldwide: foci in mountainous Mexico, Central and South America, central and eastern Africa	1–2 weeks: fever, chills, fatigue, malaise, headache, myalgia, trunk rash	N/A
	Trench fever	Bacteria *Bartonella quintana*	*Pediculus humanus humanus*	Worldwide	7–30 days: fever, fatigue, malaise, headache, rash (sometimes), tenderness, pain (shins),	N/A
	Relapsing fever (epidemic)	Bacteria *Borrelia recurrentis*	*Pediculus humanus humanus*	Worldwide	5–15 days: 1–10 relapsing fevers lasting 2–9 days each then a 2–4 day afebrile period, malaise, fatigue, rash	N/A
Fleas Order Siponaptera	Flea-borne typhus (endemic)	Bacteria *Rickettsia typhi*	Various species	Worldwide	1–2 weeks: similar to louse-borne typhus but milder	N/A

Vector group	Disease	Pathogen (type and species)	Major vector(s)	Distribution	Incubation period and initial symptoms	V, C, L²
Fleas Order Siponaptera (Continued)	Plague	Bacteria *Yersinia pestis*	*Xenopsylla* spp.: *X. cheopis*, other fleas?	Worldwide: endemic foci in USA (west of the 100th meridian), Peru, Bolivia, Brazil, Ecuador, India, Russia, five African and six Asian countries	1–7 days: fever, chills, malaise, headache, sore throat, restlessness, confusion, fatigue, shock	V/L
Cockroaches families Blattidae, Blattellidae, Blaberidae	Poliomyelitis	Virus family Enteroviridae	Various species	Limited foci in Indian subcontinent, western and central Africa	7–14 days: fever, chills, malaise, headache, nausea, vomiting, fatigue, then severe myalgia, stiff neck and flaccid paralysis in 3–4 days	V
	Bacterial food poisoning, gastroenteritis, wound and skin infections, conjunctivitis	Various bacteria, molds, and fungi	Various species	Worldwide	Various	N/A
	Amebiasis	Protozoan *Entamoeba histolytica*	Various species	Worldwide	Days to months: mild abdominal discomfort to acute, fulminant dysentery with fever, chills, bloody/ mucous diarrhoea, intestinal ulcers	N/A
	Helminthic infections	Helminths *Ancylostroma duodenale* *Ascaris* spp. *Enterobius vermicularis* *Hymenolepis* spp. *Necator americanus* *Trichuris trichuria*	Various species	Worldwide	Various	N/A
Mites	Scrub typhus	Bacteria *Orientia tsutsugamushi*	*Leptotrombidium* spp.	Central and Southeast Asia (especially northern Thailand), western Pacific	10–12 days: fever, headache, conjunctival congestion, pain, lymphadenopathy, rash after 1 week	C

Vector group	Disease	Pathogen (type and species)	Major vector(s)	Distribution	Incubation period and initial symptoms	V, C, L[2]
Mites (Continued)	Rickettsial pox	Bacteria *Rickettsia akari*	*Liponyssoides sanguineus*	Eastern USA (primarily New York), Eurasia	10–24 days: skin lesion, then in 1 week: fever, chills, myalgia, lymphadenopathy, headache, profuse sweating, then rash (except palms and soles)	N/A
	Mite dermatitis	Larval mites invade skin	Various species examples: *Cheyletiella blakei* *C. yasgursi* *Dermanyssus gallinae* *Ornithonyssus bacoti* *O. bursa* *O. sylviarum*	Worldwide	Within hours: itchy rash, typically where clothing adheres closely to body	N/A
	Scabies	Parasitic mite larvae, nymphs and adults	*Sarcoptes scabiei*	Worldwide	2–6 weeks first exposure, 1–4 days after re-exposure: pruritus	N/A
Ticks family Argasidae	Relapsing fever (tick-borne spirochetetosis)	Bacteria *Borrelia turicata* *B. duttonii* *B. hispanica* *B. persica* *B. crocidurae* *B. caucasica* *B. hermsii* *B. turicata* *B. parkeri* *B. venezuelensis* *B. mazzottii*	*Ornithodorus erraticus* *O. erraticus sonrai* *O. hermsi* *O. marocanus* *O. maubata* *O. parkeri* *O. rudis* *O. talaje* *O. tholozani* *O. turicata* *O. verrucosus*	Worldwide but foci in: Africa, Near/Far East, Central Asia, Eastern Europe, Mediterranean, western USA and Canada to Central and South America, Uzbekistan and Kashmir to Cyprus and Tripoli, Caucasus	5–15 days: 1–10 relapsing fevers lasting 2–9 days each then a 2–4 day afebrile period, malaise, fatigue, rash	N/A
	Hughes, Punte Salinas and Zirqa	Virus family Bunyaviridae	*Ornithodorus* spp.	Ethiopia, Arabian Gulf, Americas	4–5 days: fever, malaise	N/A

Vector group	Disease	Pathogen (type and species)	Major vector(s)	Distribution	Incubation period and initial symptoms	V, C, L[2]
Ticks family Argasidae (Continued)	Kyasanur Forest disease	Virus family Flaviviridae	Haemaphysalis spinigera Isolations from: Ornithodoros chiropterphila	India	3–8 days: biphasic fever, chills, malaise, headache, back/limb pain, fatigue, conjunctivitis, diarrhoea, vomiting, confusion, encephalitis, haemorrhage	N/A
	Q fever	Bacteria Coxiella burnetti	Argas spp.	Worldwide	2–3 weeks: fever, chills, malaise, headache, severe sweats, fatigue	N/A
	Quaranfil	Virus not classified	Argas spp.	Africa, Afghanistan, Nepal	4–5 days: biphasic fever, malaise	N/A
family Ixodidae	Lyme disease	Bacteria Borrelia burgdorferi Other Borrelia? B. garinii B. afzeli	Ixodes scapularis I. pacificus I. ricinus I. persulcatus	North America, Europe, Asia, Australia	3–32 days: skin lesion progressing to wheel rash, fever, malaise, headache, stiff neck, myalgia, fatigue, arthralgia, lymphadenopathy	N/A
	Bhanja	Virus family Bunyaviridae	Haemaphysalis intermedia H. punctata Hyalomma truncatum Amblyomma variegatum Boophilus decoloratus	India, Italy, Nigeria, Cameroon	4–5 days: fever, malaise	N/A
	Human granulocytic ehrlichiosis	Bacteria Ehrlichia equi	Ixodes scapularis I. ricinus	USA, Asia, Europe	7–21 days: fever, malaise, headache, anorexia, fatigue, myalgia, nausea, vomiting	N/A
	Human monocytic ehrlichiosis and anaplasmosis	Bacteria Ehrlichia chaffeensis E. ewingii E. equi E. phagocytophila	Amblyomma americanum	Southeast and southcentral USA	7–21 days: fever, malaise, headache, anorexia, fatigue, myalgia, nausea, vomiting	N/A

Vector group	Disease	Pathogen (type and species)	Major vector(s)	Distribution	Incubation period and initial symptoms	V, C, L[2]
family Ixodidae (Continued)	Tickborne typhus (Rocky Mountain spotted fever, Mexican spotted fever, Tobia fever, Sao Paulo fever)	Bacteria *Rickettsia rickettsia* *Rickettsia sibirica*	*Dermacentor andersoni* *D. marginatus* *D. occidentalis* *D. parumapertus* *D. variabilis* *Amblyomma cajennense* *Rhiphcephalus sanguineus* *Haemaphysalis* spp.	Americas (Canada, Mexico, parts of South America)	3–14 days: high fever (2–3 weeks), chills, malaise, severe headache, fatigue, myalgia, rash (including palms and soles)	N/A
	North Asian (Siberian) tick typhus	Bacteria *Rickettsia sibirica*	*Ixodes granulatus* *Rhipicephalus sanguineus* *Hyalomma asiaticum* *Dermacentor* spp. *Haemaphysalis* spp.	Eurasia	3–14 days: high fever (2–3 weeks), chills, malaise, severe headache, fatigue, myalgia, rash (including palms and soles)	N/A
	Human babesiosis	Protozoan *Babesia microti* *B. major* *B. divergens*	*Ixodes scapularis* *I. ricinus* *I. granulatus*	USA (primarily Nantucket island), Europe, Taiwan?	1–8 weeks: fever, chills, malaise, fatigue, myalgia, jaundice	N/A
	Ganjam virus	Virus family Bunyaviridae	*Haemaphysalis intermedia* *H. wellingtoni*	Indian subcontinent	4–5 days: fever, malaise	N/A
	Colorado tick fever	Virus family Reoviridae	*Dermacentor andersoni* *D. occidentalis*	USA Rocky Mountain states, Black Hills of South Dakota, British Columbia, western Canada	4–5 days: biphasic fever, malaise	N/A
	Crimean-Congo haemorrhagic fever	Virus family Bunyaviridae	*Hyalomma anatolicum* *H. scupense* *H. marginatum marginatum* *Argas reflexus* *Rhipicephalus sanguineus*	Eurasia, North Africa, Balkans, Middle East, Asia	1–3 days: fever, malaise, fatigue, myalgia, arthralgia, anorexia, haemorrhage	N/A

Vector group	Disease	Pathogen (type and species)	Major vector(s)	Distribution	Incubation period and initial symptoms	V, C, L²
family Ixodidae (Continued)	Dugbe	Virus family Bunyaviridae	Hyalomma rufipes Amblyomma variegatum A. lepidum Boophilus decoloratus	Nigeria	4–5 days: biphasic fever, malaise	N/A
	Lymphocytic choriomeningitis	Virus family Arenaviridae	Amblyomma variegatum Rhipicehalus sanguineus Dermacentor andersoni	Ethiopia, Canada	8–13 days: fever, malaise, myalgia, headache, fatigue, stiff neck	N/A
	Nairobi sheep disease	Virus family Bunyaviridae	Rhipicephalus appendiculatus Amblyomma variegatum	Eastern Africa, Somalia	4–5 days: fever, malaise	N/A
	Powassan fever	Virus family Flaviviridae	Ixodes spp. Dermacentor andersoni Haemaphysalis longicornis Ixodes spinipalpis	North America (Ontario and Quebec, Canada, New York State, Colorado, South Dakota), Russia (Maritime Province, Soviet Union)	7–14 days: fever (sometimes biphasic), chills, malaise, severe headache, myalgia, fatigue, retro-orbital pain, nausea, vomiting	N/A
	Tick-borne encephalitis (including Russian spring-summer encephalitis)	Virus family Flaviviridae	Ixodes ricinus I. persulcatus I. pavlovskyi I. hexagonus Haemaphysalis inermis H. punctata H. concinna Dermacenter marginatus D. reticulatus	Eurasia	7–14 days: fever (sometimes biphasic), chills, malaise, severe headache, myalgia, fatigue, retro-orbital pain, nausea, vomiting	V/L
	Omsk haemorrhagic fever	Virus family Flaviviridae	Dermacentor reticulatus D. marginatus D. pictus Ixodes apranophorus I. persulcatus	Eurasia	3–8 days: fever, chills, malaise, headache, back/limb pain, fatigue, conjunctivitis, diarrhoea, vomiting, haemorrhage	N/A

Vector group	Disease	Pathogen (type and species)	Major vector(s)	Distribution	Incubation period and initial symptoms	V, C, L[2]
family Ixodidae (Continued)	Kyasanur Forest disease	Virus family Flaviviridae	**Primarily:** *Haemaphysalis turturis* *H. spinigera* *Haemaphysalis* spp. *Ixodes ceylonicus* *I. petauristae* **Isolations from:** *Dermacentor auratus* *Rhipicephalus turanicus* *Ornithodoros chiropterphila*	India	3–8 days: fever, chills, malaise, headache, back/limb pain, fatigue, conjunctivitis, diarrhoea, vomiting, confusion, encephalitis, haemorrhage	N/A
	Kemerovo	Virus family Reoviridae	*Ixodes persulcatus*	Siberia	4–5 days: biphasic fever, malaise	N/A
	Langat	Virus family Flaviviridae	*Ixodes granulatus*	Malaysia, Japan	3–8 days: fever, chills, malaise, headache, fatigue	N/A
	Louping-ill	Virus family Flaviviridae	*Ixodes ricinus*	Europe, United Kingdom	7–14 days: biphasic fever, chills, malaise, myalgia, severe headache, fatigue, retro–orbital pain, nausea, vomiting	N/A
	Lymphocytic choriomeningitis	Virus family Arenaviridae	*Amblyomma variegatum* *Rhipicephalus sanguineus* *Dermacentor andersoni*	Ethiopia, Canada	8–13 days: fever, malaise, headache, myalgia, fatigue stiff neck	N/A
	Heartland virus	Virus family Phenuiviridae	*Amblyomma americanum*	Central United States	<2 weeks: fever, weakness, headache, myalgia, fatigue, diarrhoea, weight loss, bruising, arthralgia, leukopenia	N/A
	Bourbon virus	Virus family Orthomyxoviridae	*Amblyomma americanum*	Central United States	<21 days: fever, headache, myalgia, fatigue, nausea, vomiting, diarrhoea, rash on chest, abdomen and back, respiratory distress syndrome, organ failure	N/A

Vector group	Disease	Pathogen (type and species)	Major vector(s)	Distribution	Incubation period and initial symptoms	V, C, L²
family Ixodidae (Continued)	Queensland tick typhus	Bacteria *Rickettsia australis*	*Ixodes holocyclus*	Australia, Tasmania	7–10 days: skin lesion, fever, malaise, palm and sole rash, lymphadenopathy	N/A
	African tick-bite fever (African spotted fever)	Bacteria *Rickettsia africae*	*Amblyomma hebraeum* *A. variegatum*	Africa	1–15 days: skin lesion, fever, malaise, oedema, lymphadenopathy	N/A
	Thogoto meningoencephalitis	Virus family Orthomyxoviridae	*Hyalomma anatolicum* *Boophilus decoloratus* *H. truncatum* *Amblyomma variegatum* *Rhipicephalus bursa*	Egypt, Kenya, Nigeria, Sicily	4–5 days: fever, malaise, fatigue, meningitis	N/A
	Oriental spotted fever	Bacteria *Rickettsia japonica*	*Dermacentor taiwanensis* *Haemaphysalis flava*	Japan	3–14 days: fever, malaise, rash	N/A
	Boutonneuse fever (Mediterranean spotted fever, South African tick typhus, Kenya tick typhus, Crimean tick typhus, Marseilles fever, Indian tick typhus, Astrakhan fever)	Bacteria *Rickettsia conorii* and others: *R. slovaca* *R. helvetica* *R. mongolotimonae*	*Amblyomma* spp. *Dermacentor* spp. *Haemaphysalis leachi* *Hyaloma rufipes* *Ixodes* spp. *Rhipecephalus sanguineus* *R. pumilio*	Southern Europe, Crimea, Russia, Africa, India, Middle East, southern and western Asia	5–7 days: skin lesion, fever, malaise, palm and sole rash, lymphadenopathy	N/A
	Tularemia (rabbit fever)	Bacteria *Francisella tularensis*	*Amblyomma americanum* *Dermacentor andersoni* *D. occidentalis* *D. variabilis* *Haemaphysalis leporispalustris* *Ixodes* spp. *Rhipecephalus sanguineus*	Worldwide but foci in: North America, central Europe, Russia, Japan, Israel, Africa	1–14 days: fever, chills, fatigue, headache, myalgia, arthralgia, vomiting, lymphadenopathy	N/A

Vector group	Disease	Pathogen (type and species)	Major vector(s)	Distribution	Incubation period and initial symptoms	V, C, L[2]
family Ixodidae (Continued)	Q fever	Bacteria *Coxiella burnetti*	*Amblyomma* spp. *Dermacentor* spp. *Haemaphysalis* spp. *Ixodes* spp.	Worldwide	2–3 weeks: fever, chills, malaise, headache, severe sweats, fatigue	N/A
	Wanowrie	Virus family Bunyaviridae	*Hyalomma impeltatum* *H. isaaci*	Egypt, India, Iran, Sri Lanka	4–5 days: fever, malaise, fatigue, haemorrhage	N/A
Snails	Schistosomiasis	Helminths *Schistosoma haematobium* *S. japonicum* *S. intercalatum* *S. mansoni* *S. mekongi* *S. mattheei*	*Biomphalaria* spp. *Bulinus* spp. *Ferrissia* spp. *Ocomelania* spp. *Planorbis* spp. *Tricula* spp.	Africa, Asia, Indonesia, Philippines, India, Taiwan, Portugal, Middle East, Madagascar, Caribbean	4–6 weeks: fever, chills, headache, malaise, fatigue; other symptoms vary by species and location but generally include: abdominal pain, diarrhoea, nausea, vomiting, cough, urticaria, haematuria, hepatomegaly, lymphadenopathy	N/A

1. Modified from a Guide to Entomological Surveillance during Contingency Operations, 2001, U.S. Army.
2. Per WHO, CDC and PAHO guidelines and peer-reviewed scientific literature as of December 2016.
   C = chemoprophylaxis; L = limited availability or access, often restricted to use in highly endemic areas, only in certain countries or only used during high-risk activities/occupations; V = vaccine

Index

paralysis 33, 38, 39–40
parasites 67
  bot flies 129
  chiggers 67
  fleas 156
  freshwater snails and 42
  kissing bugs and 100
parasitic otitis 3, 76, 80, 169
pathogens 169
  filth flies 152
  ticks 77
pavement ants (*Tetramorium caespitum*) 112
peak activity periods, disease vectors 9
pederin toxin, *Paederus* beetles 127
pedicellariae, sea urchins 32
pedipalps (scorpions) 60
Pelinobius spp. 51–2
perianal 169
permethrin 10, 11–12, 69, 96, 97
permethrin-treated bedding 102, 136
permethrin-treated clothing 17, 89, 111, 150
Persian fowl-tick (*Argas persicus*) 78, 79
personal health summary 18, 108
personal protection measures 8–18, 71, 103,
    105, 136, 152
  biting midges and 142, 143
  filth flies and 155
  fleas and 161
  mites and 72
  mosquitoes and 137
  scorpion stings and 65–6
  spider bite and 57
  tick-borne disease 75, 88–9, 90
Peru 157
pesticides 10, 17
pest management professionals 57, 105, 162
pet fur infestations, rabbit fur mite and 73
pet washing, fleas and 162
petroleum jelly, as wound treatment 137
Pharaoh ant (*Monomorium pharaonis*) 112
pheromones 169
  bee stings 114
  hymenopterans 107
Philippines 43
Phlebotominae 143–4
pigeon tick (*Argus reflexus*) 78–9

piloerection 169
pink jellyfish (*Pelagia noctiluca*) 25
pin-prick bites, plant-feeding bugs 106
plague 4, 157, 159, 160, 201
plague bacteria (*Yersinia pestis*), fleas
    and 157, 159, 160
pneumatophore 26
pneumonia 150
point source threats, invertebrates 5
poliomyelitis 76
Polyceroconas spp. 93
Polychaetes (Class Polychaeta) 36–7
polypeptides, scorpion venom 65
Portuguese man o'war (*Physalia physalis*)
    26–7
  as prey 41
post-inflammatory pigmentation 25
Powossan encephalitis virus 84, 85, 202
Praziquantel 43
pregnancy, disease vectors and 16, 18
prevention and treatment
  biting fly attacks 150–1
  eye gnat transmitted diseases 152
  filth fly infestation 155
  flea infestation 161–2
  scorpion stings 65–6
  spider bite 57–8
  tick-borne disease 88–91
primaquine 8
prophylactic medication 17, 18, 90
prostration 45, 169
protective clothing 12–14, 57, 66, 128
  biting flies and 150
  ticks and 89
Psorophora spp. 137, 142
psychological threats, invertebrates 5, 6–7
pteropods 39
pubic louse (crab louse) (*Pthirus pubis*) 97
pupal stage
  fleas 156–7
  mosquitoes 138
  screwworm flies 133
puss caterpillar (*Megalopyge opercularis*) 123,
    124
pustules 42, 69, 169
  caterpillar contact 122